P9-ELD-634

# Programming
# Arduino™ Next Steps

CALGARY PUBLIC LIBRARY

JAN -    2014

## About the Author

**Dr. Simon Monk** (Preston, UK) has a degree in cybernetics and computer science and a PhD in software engineering. Dr. Monk spent several years as an academic before he returned to industry, co-founding the mobile software company Momote Ltd. He has been an active electronics hobbyist since his early teens and is a full-time writer on hobby electronics and open source hardware. Dr. Monk is the author of numerous electronics books, specializing in open source hardware platforms, especially Arduino and Raspberry Pi. He is also co-author with Paul Scherz of *Practical Electronics for Inventors, Third Edition*.

You can follow Simon on Twitter, where he is @simonmonk2.

# Programming
# Arduino™ Next Steps
## Going Further with Sketches

**Simon Monk**

New York   Chicago   San Francisco   Athens
London   Madrid   Mexico City   Milan
New Delhi   Singapore   Sydney   Toronto

Library of Congress Cataloging-in-Publication Data

Monk, Simon.
   Programming Arduino next steps : going further with sketches / Dr.
Simon Monk.
      pages cm
   ISBN 978-0-07-183025-6 (pbk.)
   1. Arduino (Programmable controller)—Programming.   2. Programmable
controllers—Programming.   I. Title.
   TJ223.P76M6653 2014
   629.8'9551—dc23

                                                              2013033821

McGraw-Hill Education books are available at special quantity discounts to use as premiums and sales promotions, or for use in corporate training programs. To contact a representative, please visit the Contact Us pages at www.mhprofessional.com.

**Programming Arduino™ Next Steps: Going Further with Sketches**

Copyright © 2014 by McGraw-Hill Education. All rights reserved. Printed in the United States of America. Except as permitted under the Copyright Act of 1976, no part of this publication may be reproduced or distributed in any form or by any means, or stored in a database or retrieval system, without the prior written permission of publisher, with the exception that the program listings may be entered, stored, and executed in a computer system, but they may not be reproduced for publication.

All trademarks or copyrights mentioned herein are the possession of their respective owners and McGraw-Hill Education makes no claim of ownership by the mention of products that contain these marks.

1 2 3 4 5 6 7 8 9 0    DOC DOC    1 0 9 8 7 6 5 4 3

ISBN      978-0-07-183025-6

MHID      0-07-183025-1

| | | |
|---|---|---|
| **Sponsoring Editor** | **Copy Editor** | **Composition** |
| Roger Stewart | LeeAnn Pickrell | Cenveo Publisher |
| **Editorial Supervisor** | **Proofreader** | Services |
| Jody McKenzie | Claire Splan | **Illustration** |
| **Project Manager** | **Indexer** | Cenveo Publisher |
| Vastavikta Sharma, | James Minkin | Services |
| Cenveo® Publisher | **Production Supervisor** | **Art Director, Cover** |
| Services | James Kussow | Jeff Weeks |
| **Acquisitions Coordinator** | | |
| Amy Stonebraker | | |

Information has been obtained by McGraw-Hill Education from sources believed to be reliable. However, because of the possibility of human or mechanical error by our sources, McGraw-Hill Education, or others, McGraw-Hill Education does not guarantee the accuracy, adequacy, or completeness of any information and is not responsible for any errors or omissions or the results obtained from the use of such information.

# CONTENTS AT A GLANCE

# CONTENTS

# ACKNOWLEDGMENTS

**Many thanks** to all those at McGraw-Hill Education who have done such a great job in producing this book. In particular, thanks to my editor Roger Stewart and to Vastavikta Sharma, Jody McKenzie, LeeAnn Pickrell, and Claire Splan.

I would also like to thank Adafruit, SparkFun, and CPC for supplying many of the modules and components used in the preparation of this book.

And last but not least, thanks once again to Linda, for her patience and generosity in giving me space to do this.

# INTRODUCTION

**Arduino has** become the standard microcontroller used by makers, artists, and educators due to its ease of use, low cost, and plethora of interface boards (shields). Plug-in shields can be attached to the basic board, extending the Arduino into the Internet, robotic, and home automation realms.

Simple Arduino projects are easy to make. As soon as you start to stray into territory not covered by the introductory texts, however, you'll find that things can rapidly become confusing and frustrating as complexity—the enemy of all programmers—rears its ugly head.

This book is designed as a companion and sequel to the very successful book *Programming Arduino: Getting Started with Sketches*. Although this book includes a brief recap of basic Arduino Programming, it leads the reader through the more advanced aspects of Arduino programming. Specifically, this book will help you with:

- Working effectively with minimal memory
- Doing more than one thing at a time, without the luxury of multithreading
- Packaging your code in libraries for others to use
- Using hardware and timer interrupts
- Maximizing performance
- Minimizing power consumption
- Interfacing with different types of serial busses (I2C, 1-Wire, SPI, and serial)
- USB programming
- Network programming
- Digital Signal Processing (DSP)

# Downloads

The book includes some 75 example sketches, which are all open source and available on the author's website at www.simonmonk.org. Follow the link to the pages for this book where you will be able to download the code as well as an up-to-date list of errata for the book.

# What Will I Need?

This book is primarily about software. So, for most of the examples, all you really need is an Arduino and an LED or multimeter. Having said that, if you do have other Arduino shields, these will come in handy. You will also need an Ethernet or Wi-Fi shield for Chapter 12. Throughout the book, several different types of module are used to illustrate different interfaces.

Although the book is mostly concerned with the Arduino Uno (the most commonly used Arduino board), it also covers some of the special features of other Arduino boards like the Leonardo and Arduino Due for USB programming and Digital Signal Processing.

The Appendix at the end of this book lists possible suppliers for these parts.

# Using This Book

Each of the chapters deals with a specific topic relating to Arduino programming. Apart from Chapter 1, which is a recap and overview of Arduino basics, the remaining chapters can be accessed pretty much in any order you like.

If you are an experienced developer in other areas, then you might like to read Chapter 14 first to put Arduino programming into context.

Following is a description of each chapter:

1. **"Programming Arduino"**   This chapter contains a summary of Arduino programming. It is a primer for those needing to get up to speed quickly with basic Arduino.

2. **"Under the Hood"**   In this chapter, we take a peek under the hood at how the Arduino software works and where it came from.

3. **"Interrupts and Timers"**   Novices often steer clear of using interrupts. They shouldn't, however, as they can be handy on occasion and are not difficult to code for. Although there are some pitfalls, this chapter tells you what you need to aware of.

4. **"Making Arduino Faster"**   Arduinos have low-speed, low-power processors and sometimes you need to squeeze every ounce of juice out of them. For example, the built-in **digitalWrite** function is safe and easy to use, but is not very efficient, especially when setting multiple outputs at the same time. In this chapter, you look at ways to exceed this performance and learn about other techniques for writing time-efficient sketches.

5. **"Low Power Arduino"**   When you want to run your Arduino on batteries or solar, then you need to look at minimizing power consumption. In addition to optimizing the hardware design, you can also set up the code to reduce the Arduino's energy use.

6. **"Memory"**   In this chapter, we look at minimizing memory usage and the benefits and dangers associated with using memory dynamically within your sketches.

7. **"Using I2C"**   The Arduino's I2C interface can greatly simplify talking to modules and components, reducing the number of interface pins you need to use. This chapter describes how I2C works and how to use it.

8. **"Interfacing with 1-Wire Devices"**   This chapter focuses on 1-wire bus devices such as Dallas Semiconductor's range of temperature sensors, which are extremely popular for use with the Arduino. You learn how the bus works and how to use it.

9. **"Interfacing with SPI Devices"**   Yet another interface standard used with the Arduino is SPI. This chapter explores how it works and how to use it.

10. **"Serial UART Programming"**   Serial communications, either through USB or the Arduino's Rx and Tx pins, provide a great way to exchange data between peripherals and other Arduinos. In this chapter, you learn how to use serial.

11. **"USB Programming"**   This chapter looks at various aspects of using the Arduino with USB. You'll learn about the keyboard and mouse emulation features provided by the Arduino Leonardo and also the reverse process of allowing a USB keyboard or mouse to be connected to a suitably equipped Arduino.

12. **"Network Programming"**   The Arduino is a common component in the Internet of Things. In this chapter, you'll learn how to program the Arduino for the Internet. Topics include Wi-Fi and Ethernet shields as well as using web services and the Arduino as a mini web server.

13. **"Digital Signal Processing"**   The Arduino is capable of fairly rudimentary signal processing. This chapter discusses a variety of techniques, from filtering a signal from an analog input using software rather than external electronics to calculating the relative magnitude of various frequencies in a signal using the Fast Fourier Transform.

14. **"Managing with One Process"**   Programmers coming to Arduino from a background of programming large systems often signal the lack of multithreading and concurrency in Arduino as some kind of deficiency. In this chapter, I try to set the record straight and show how to embrace the single-thread model of embedded systems.

15. **"Writing Libraries"**   Sooner or later, you will make something really good that you think other people could use. This is the time to wrap up the code in a library and release it to the world. This chapter shows you how.

# Resources

This book is supported by accompanying pages on the author's website (www.simonmonk.org). Follow the link for this book, and you will find all the source code, as well as other resources such as errata.

# 1

# Programming Arduino

This chapter summarizes the basics of Arduino. If you are completely new to Arduino, then you might find it useful to also read *Programming Arduino: Getting Started with Sketches* (McGraw-Hill Professional, 2012).

## What Is Arduino?

The term *Arduino* is used to describe both the physical Arduino board (of which the most popular type is the Arduino Uno) and the Arduino system as a whole. The system also includes the software you need to run on your computer (to program the board) and the peripheral shields that you can plug into an Arduino board.

To use an Arduino, you also need a "proper" computer. This can be a Mac, Windows PC, Linux PC, or even something as humble as a Raspberry Pi. The main reason that you need the computer is so you can download programs onto the Arduino board. Once installed on the Arduino, these programs can then run independently.

Figure 1-1 shows an Arduino Uno.

The Arduino can also communicate with your computer over USB. While the computer is connected, you can send messages in both directions. Figure 1-2 shows the relationship between the Arduino and your computer.

An Arduino is unlike a conventional computer in that it has hardly any memory, no operating system, and no keyboard mouse or screen interface.

**Figure 1-1**    *An Arduino Uno*

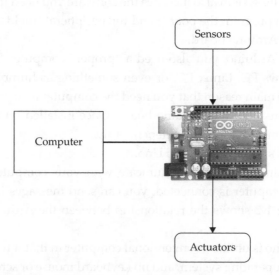

**Figure 1-2**    *The Arduino and your computer*

Its purpose is to control things by interfacing with sensors and actuators. So, for instance, you might attach a sensor to measure the temperature and a relay to control the power to a heater.

Figure 1-3 shows some of the things that you can attach to an Arduino board. There are no doubt many more types of devices that you can connect to an Arduino board.

Here is a short selection of some of the amazing projects that have been built using an Arduino:

- Bubblino—an Arduino linked to a bubble machine that blows bubbles when you tweet it!
- 3D LED cubes

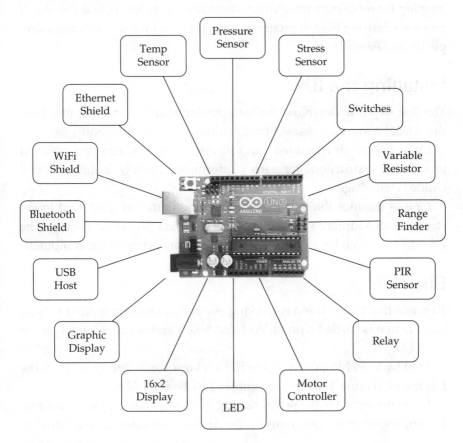

**Figure 1-3**  *Interfacing with an Arduino*

- Geiger counters
- Musical instruments
- Remote sensors
- Robots

# Installation and the IDE

The software that you use to program the Arduino is called the *Arduino Integrated Development Environment (IDE)*. If you are a software developer and accustomed to using complex IDEs like Eclipse or Visual Studio, you'll find the Arduino IDE very simple—and possibly find yourself wishing for repository integration, command completion, and the like. If you are relatively new to programming, you will love the Arduino's simplicity and ease of use.

## Installing the IDE

The first step is to download the software for your type of computer from the official Arduino website: http://arduino.cc/en/Main/Software.

Once you've downloaded the software, then you can find detailed installation instructions for each platform here: http://arduino.cc/en/Guide/HomePage.

One of the nice things about the Arduino is that all you need to get started is an Arduino, a computer, and a USB lead to connect the two. The Arduino can even be powered over the USB connection to the computer.

## Blink

To prove that the Arduino is working, we are going to program it to flash an LED that is labeled *L* on the Arduino board and hence is known as the "L" LED.

Start by launching the Arduino IDE on your computer. Then, from the File menu, (Figure 1-4) select Examples | 01 Basics | Blink.

In an attempt to make programming the Arduino sound less daunting to nonprogrammers, programs on the Arduino are referred to as *sketches*.

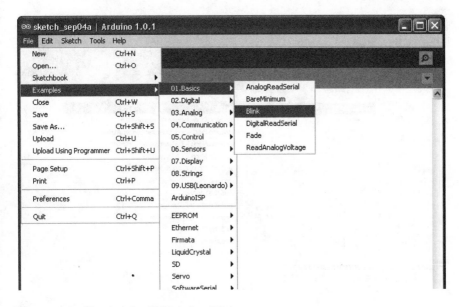

**Figure 1-4**   *The Arduino IDE loading Blink*

Before you can send the Blink sketch to your Arduino, you need to tell the Arduino IDE what type of Arduino you're using. The most common type is the Arduino Uno, and in this chapter, I assume that is what you have. So from the Tools | Board menu, select Arduino Uno (Figure 1-5).

As well as selecting the board type, you also need to select the port it is connected to. In Windows this is easy, as it is usually COM4 and will probably be the only port in the list (see Figure 1-6). On a Mac or Linux computer, however, there will generally be more serial devices listed. The Arduino IDE shows the most recently connected devices first, so your Arduino board should be at the top of the list.

To actually upload the sketch onto the Arduino board, click the Upload button on the toolbar. This is the second button on the toolbar, which is highlighted in Figure 1-7.

Once, you click the Upload button, a few things should happen. First, a progress bar will appear as the Arduino IDE compiles the sketch (meaning it converts the sketch into a suitable form for uploading). Then, the LEDs

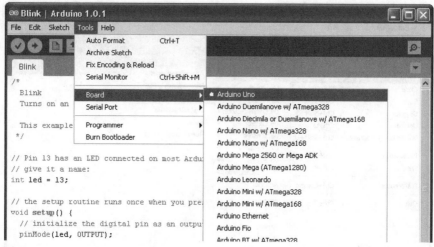

**Figure 1-5**   *Selecting the board type*

on the Arduino labeled *Rx* and *Tx* should flicker for a while. Finally, the LED labeled *L* should start to blink. The Arduino IDE will also display a message like "Binary sketch size: 1,084 bytes (of a 32,256 byte maximum)." This means the sketch has used about 1kB of the 32kB of the flash memory available for programs on the Arduino.

Before you start programming, let's have a look at the hardware that your programs, or sketches, will have to work within and have available for their use.

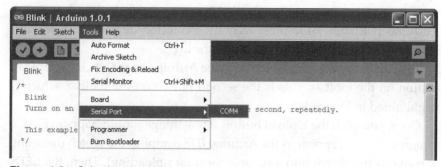

**Figure 1-6**   *Selecting the serial port*

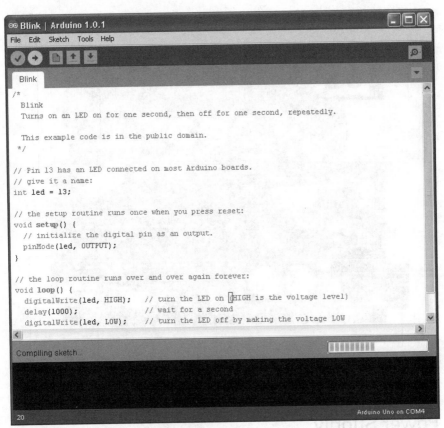

**Figure 1-7**  *Uploading the Blink sketch*

# A Tour of Arduino

Figure 1-8 shows the anatomy of an Arduino Board. Starting at the top, next to the USB socket in the top-left corner, is the Reset switch. Clicking this sends a logic pulse to the microcontroller's Reset pin, clearing the microcontroller's memory so it can start its program fresh. Note that any program stored on the device is retained because it is kept in *nonvolatile* flash memory—that is, memory that remembers even when the device is not powered on.

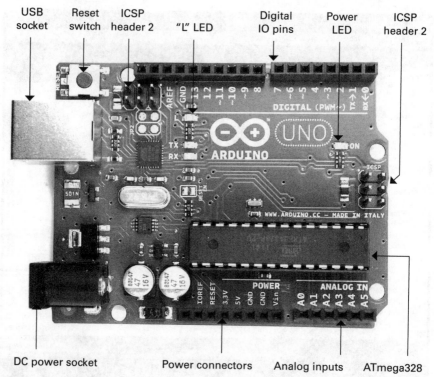

**Figure 1-8**   *Anatomy of an Arduino board*

## Power Supply

The Arduino can either be powered through either the USB connection or the DC power socket below it. When powering the Arduino from a DC adaptor or batteries, anything between 7.5 and 12V DC can be supplied through the power socket. The Arduino itself only uses about 50mA. So a small PP3 9V battery (200mAh) will power it for around 40 hours.

When the Arduino is powered on, the power LED on the right of the Uno (on the left of the Leonardo) is lit.

## Power Connections

Next, let's look at the connectors at the bottom of Figure 1-8. Apart from the first connection, you can read the connection names next to the connectors.

The first unlabeled connection is reserved for later use. The next pin, *IOREF,* indicates the voltage at which the Arduino operates. Both the Uno and Leonardo operate at 5V, so this pin will always be set at 5V, but you will not use it for anything described in this book. Its purpose is to allow shields attached to 3V Arduinos like the Arduino Due to detect the voltage at which the Arduino operates.

The next connect is Reset. This connection does the same thing as pressing the Reset switch on the Arduino. Rather like rebooting a PC, it resets the microcontroller to begin its program from the start. The Reset connector allows you to reset the microcontroller by momentarily setting this pin low (connecting it to GND). It is fairly unlikely that you'll need to do this, but it's quite nice to know that the connector is there.

The remaining pins provide different voltages (3.3, 5, GND, and 9), as labeled. *GND,* or *ground,* just means zero volts. It is the reference voltage to which all other voltages on the board are relative.

The two GND connections are identical; having more than one GND pin to connect things to is useful. In fact, there is another GND socket at the top of the board.

## Analog Inputs

The next section of connections is labeled Analog In 0 to 5. These six pins can be used to measure the voltage connected to them so the value can be used in a sketch. Although labeled as analog inputs, these connections can also be used as digital inputs or outputs. By default, however, they are analog inputs.

## Digital Connections

Now let's switch to the top connector, starting on the right side (Figure 1-8). We have pins labeled Digital 0 to 13. These can be used as either inputs or outputs. When using them as outputs, you can control them from a sketch. If you turn them on from your sketch, they will be at 5V, and if you turn them off, they will be at 0V. As with the supply connectors, you have to be careful not to exceed their maximum current capabilities.

These connections can supply 40mA at 5V—more than enough power to light a standard LED, but not enough to drive an electric motor directly.

# Arduino Boards

The Arduino Uno (Figure 1-1) is the current incarnation of the original Arduino board. It is the most common Arduino board and is generally what people mean when they say they are using an Arduino.

The other types of Arduino board all satisfy special requirements, like the need for more I/O (input/output) connections, faster performance, or a smaller board, or to be stitched into clothing, connect to Android phones, or integrate easily with wireless and so on.

No matter how different the hardware, each board is programmed from the Arduino IDE, with only minor variations in the software features they can use. Once you have learned how to use one Arduino Board, you have pretty much learned how to use all of them.

Let's look at the current range of official Arduino boards. There are other Arduinos than the ones discussed here, but they tend not to be that popular. For a full list of boards, check out the official Arduino website (www.arduino.cc).

## Uno and Similar

The Uno R3 is the latest of a series of "standard" boards that include the plain Uno, Duemilanove, Diecimila, and NG. These boards all use the ATmega168 or ATmega328 microprocessors, which are pretty much the same, apart from differing amounts of memory.

The other current Arduino, with the same size and connections as the Uno R3, is the Arduino Leonardo (Figure 1-9). As you can see, the board is much more sparsely populated than the Uno. This is because it uses a different processor. The Leonardo uses the ATmega32u4, which is similar to the ATmega328 but includes a built-in USB interface, removing the need for the extra components that you find on the Uno. Moreover, the Leonardo has slightly more memory, more analog inputs, and other benefits. It is also less expensive than the Uno. In many respects, it is also a better design than the Uno.

**Figure 1-9**  *The Arduino Leonardo*

If this is the case, then you might be wondering why the Leonardo is not the most popular Arduino board, rather than the Uno. The reason is that the improvements offered by the Leonardo come at the cost of making it slightly incompatible with the Uno and its predecessors. Some expansion shields (especially old designs) will not work on the Leonardo. In time, these differences will become less of a problem. At that point, it will be interesting to see if the Leonardo and its successors become the more popular boards.

The Arduino Ethernet is a relatively new addition to the Arduino stable. It combines basic Uno features with an Ethernet interface, allowing you to connect it to a network, without having to add an Ethernet shield.

## Big Arduino Boards

Sometimes an Uno or Leonardo just doesn't have enough I/O pins for the application that you intend to use it for. The choice then arises of either using hardware expansion for the Uno or switching to a bigger board.

**TIP**   *If you are coming to Arduino for the first time, do not buy one of these larger boards. It is tempting because they are bigger and faster, but they have shield compatibility problems and you will be much better off with a "standard" Uno.*

**Figure 1-10**   *The Arduino Due*

The super-sized Arduinos have the same sockets as an Uno, but then they add a double row of extra I/O pins on the end and a longer length of pins along the side (Figure 1-10).

Traditionally, the "bigger" board would be an Arduino Mega 2560. These boards, in common with all the larger Arduino boards, have more of every kind of memory. The Mega 2560 and Mega ADK both use processors with similar power to the Arduino Uno. However, the Arduino Due is an altogether more powerful beast. This power comes in the form of a 84 MHz processor (compared with the Uno's 16 MHz) but at the cost of further compatibility problems. The biggest of these is that the Due operates at 3.3V rather than the 5Vs of most previous Arduinos. Not surprisingly, this means that many Arduino shields are incompatible with it.

For the most demanding projects, however, this board has many advantages.

- Lots of memory for programming and data
- Hardware music output capabilities (hardware digital to analog converters)
- Four serial ports
- Two USB ports
- USB host and OTG interfaces
- USB keyboard and mouse emulation

## Small Arduino Boards

Just as the Uno is too small for some projects, it can also be too big for others. Although Arduino boards are low cost, it gets expensive if you start leaving one embedded in every project you make. There are a range of smaller and "pro" Arduino boards, designed either to be physically smaller than a regular Uno or to keep costs down by omitting features not required in most projects.

Figure 1-11 shows an Arduino Mini. These boards do not have a USB interface; rather, you need a separate adaptor module to program them. As well as the Mini, there are also Nanos and Micros, both of which have built-in USB but cost more.

## LilyPad and LilyPad USB Boards

One of the most interesting Arduino styles is the LilyPad (Figure 1-12) and the newer LilyPad USB. These boards are designed to be stitched into clothing using conductive threads and a range of similar LilyPad modules—for LEDs, switches, accelerometers, and so on. The older LilyPad boards require a separate USB interface, the same one required for the Arduino Mini. However, these boards are gradually being replaced by the Arduino LilyPad USB, which has a built-in USB connector.

## Unofficial Arduinos

As well as the "official" boards just described, there are also many unofficial copies and variations on the Arduino hardware, given its open

**Figure 1-11**  *An Arduino Mini and Programmer*

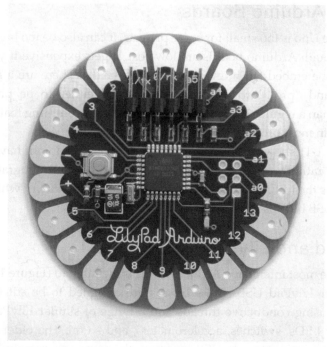

**Figure 1-12**  *An Arduino LilyPad*

source status. Straight Arduino clones are easy to come by on eBay and other low-cost outlets and are simply copies of the Arduino designs. They are only really of interest because of their price. There are also some interesting Arduino-based designs that offer extra features.

Two examples of these kind of boards that are worth looking at are

- **EtherTen**    Arduino Ethernet-type board (www.freetronics.com/products/etherten)

- **Leostick**    A slim-line Leonardo-type board with built-in USB plug (www.freetronics.com/collections/arduino/products/leostick)

Now that you have a bit more information about the hardware side of an Arduino, we can turn to programming it.

# Programming Language

A common misconception about Arduinos is that they have their own programming language. Actually, they are programmed in the language simply called C. This language has been around since the early days of computing. What Arduino does provide is a nice easy-to-use set of commands written in C that you can use in your programs.

Purists may wish to point out that Arduino uses C++, the object-oriented extension to C. Although, strictly speaking, this is true, having only 1 or 2kB of memory available generally means that the kinds of habits encouraged by object-oriented programming are not normally a good idea with Arduino, so aside from a few specialized areas, you are effectively programming in C.

Let's start by modifying the Blink sketch.

# Modifying the Blink Sketch

It may be that your Arduino was already blinking when you first plugged it in. That is because the Arduino is often shipped with the Blink sketch installed.

If this is the case, then you might like to prove to yourself that you have actually done something by changing the blink rate. Let's look at the Blink sketch to see how to change it to make the LED blink faster.

The first part of the sketch is just a comment telling you what the sketch is supposed to do. A comment is not actual program code. Part of the preparation for the code being uploaded is for all such "comments" to be stripped out. Anything between /* and */ is ignored by the computer, but should be readable by humans.

```
/*
  Blink
  Turns on an LED for one second, then off for one second, repeatedly.

  This example code is in the public domain.
*/
```

Then, there are two individual line comments, just like the block comments, except they start with //. These comments tell you what is happening. In this case, the comment helpfully tells you that pin 13 is the pin we are going to flash. We have chosen that pin because on an Arduino Uno it is connected to the built-in "L" LED.

```
// Pin 13 has an LED connected on most Arduino boards.
// give it a name:
int led = 13;
```

The next part of the sketch is the **setup** function. Every Arduino sketch must have a **setup** function, and this function runs every time the Arduino is reset, either because (as the comment says) the Reset button is pressed or the Arduino is powered up.

```
// the setup routine runs once when you press reset:
void setup() {
  // initialize the digital pin as an output.
  pinMode(led, OUTPUT);
}
```

The structure of this text is a little confusing if you are new to programming. A *function* is a section of code that has been given a name (in this case, the name is **setup**). For now, just use the previous text as a template and know that you must start your sketch with the first line **void setup() {** and then enter the commands that you want to issue, each on a line ending with a semicolon (*;*). The end of the function is marked with a } symbol.

In this case, the only command Arduino will issue is the **pinMode(led, OUTPUT)** command that, not unsurprisingly, sets that pin to be an output.

Next comes the juicy part of the sketch, the **loop** function.

Like the **setup** function, every Arduino sketch has to have a **loop** function. Unlike **setup**, which only runs once after a reset, the **loop** function runs continuously. That is, as soon as all its instructions have been run, it starts again.

In the **loop** function, you turn on the LED by issuing the **digitalWrite(led, HIGH)** instruction. You then set the sketch to pause for a second by using the command **delay(1000)**. The value 1000 is for 1000 milliseconds or 1 second. You then turn the LED back on again and delay for another second before the whole process starts over.

```
// the loop routine runs over and over again forever:
void loop() {
  digitalWrite(led, HIGH);    // turn the LED on (HIGH is the voltage level)
  delay(1000);                // wait for a second
  digitalWrite(led, LOW);     // turn the LED off by making the voltage LOW
  delay(1000);                // wait for a second
}
```

To modify this sketch to make the LED blink faster, change both occurrences of 1000 to be **200**. These changes are both in the **loop** function, so your function should now look like this:

```
void loop() {
  digitalWrite(led, HIGH);    // turn the LED on (HIGH is the voltage level)
  delay(200);                 // wait for a second
  digitalWrite(led, LOW);     // turn the LED off by making the voltage LOW
  delay(200);                 // wait for a second
}
```

If you try and save the sketch before uploading it, the Arduino IDE reminds you that it is a "read-only" example sketch, but it will offer you the option to save it as a copy, which you can then modify to your heart's content.

You do not have to do this; you can just upload the sketch unsaved. But if you do decide to save this or any other sketch, you will find that it then appears in the File | Sketchbook menu on the Arduino IDE.

So, either way, click the Upload button again, and when the uploading is complete, the Arduino resets itself and the LED should start to blink much faster.

# Variables

Variables give a name to a number. Actually, they can be a lot more powerful than this, but for now, we'll use them for this purpose.

When defining a variable in C, you have to specify the type of variable. For example, if you want your variables to be whole numbers, you would use *int* (short for *integer*). To define a variable called **delayPeriod** with a value of **200**, you need to write:

```
int delayPeriod = 200;
```

Notice that because **delayPeriod** is a name, there cannot be any spaces between words. The convention is to start variables with a lowercase letter and begin each new word with an uppercase letter. Programmers often call this *bumpy case* or *camel case*.

Let's fit this into the blink sketch, so that instead of "hard-coding" the value **200** for the length of delay, we use a variable instead:

```
int led = 13;
int delayPeriod = 200;

void setup()
{
  pinMode(led, OUTPUT);
}

void loop()
{
 digitalWrite(led, HIGH);
 delay(delayPeriod);
 digitalWrite(led, LOW);
 delay(delayPeriod);
}
```

At each place in the sketch where we used to refer to **200**, we now refer to **delayPeriod**.

Now, if you want to make the sketch blink faster, you can just change the value of **delayPeriod** in one place.

# If

Normally, your lines of program are executed in order one after the other, with no exceptions. But what if you don't want to do that? What if you only want to execute part of a sketch if some condition is true?

A good example of that might be to only do something when a button, attached to the Arduino, is pressed. The code might look like this:

```
void setup()
{
  pinMode(5, INPUT_PULLUP);
  pinMode(9, OUTPUT);
}
```

```
void loop()
{
  if (digitalRead(5) == LOW)
  {
    digitalWrite(9, HIGH);
  }
}
```

In this case, the condition (after the **if**) is that the value read from pin 5 has a value of **LOW**. The double equals symbol **==** is used for comparing two values. It is easy to confuse it with a single equals sign that assigns a value to a variable. An **if** statement says, if this condition is true, then the commands inside the curly braces are executed. In this case, the action is to set digital output to **9, HIGH**.

If the condition is not true, then the Arduino just continues on with the next thing. In this case, that is the loop function, which runs again.

# Loops

As well as conditionally performing some of the actions, you also need your sketch to be able to repeat actions over and over again. You get this for free of course by putting commands into the sketch's **loop** function. That is, after all, what happens with the Blink example.

Sometimes, however, you'll need to be more specific about the number of times that you want to repeat something. You can accomplish this with the **for** command, which allows you to use a counter variable. For example, let's write a sketch that blinks the LED ten times. Later, you'll see why this approach might be considered less than ideal under some circumstances, but for now, it will do just fine.

```
// sketch 01_01_blink_10
int ledPin = 13;
int delayPeriod = 200;
void setup()
{
  pinMode(ledPin, OUTPUT);
}
void loop()
{
```

```
for (int i = 0; i < 10; i++)
{
  digitalWrite(ledPin, HIGH);
  delay(delayPeriod);
  digitalWrite(ledPin, LOW);
  delay(delayPeriod);
}
}
```

***NOTE***   *As this is the first full sketch, it's named in a comment at the top of the file. All the sketches named in this way can be downloaded from the author's website at www.simonmonk.org.*

*To install all the sketches into your Arduino environment, unzip the file containing the sketches into your Arduino directory, which you'll find in your Documents folder. The Arduino IDE automatically creates this folder for you the first time it is run.*

The **for** command defines a variable called **i** and gives it an initial value of **0**. After the **;** the text **i < 10** appears. This is the condition for staying in the loop. In other words, while **i** is less than **10**, keep doing the things inside the curly brackets.

The last part of the **for** command is **i++**. This is C shorthand for "i = i + 1" which, not surprisingly, adds 1 to the value of **i**. One is added to the value of **i** each time around the loop. This is what ensures that you can escape from the loop, because if you keep adding 1 to **i**, eventually it will be greater than 10.

# Functions

Functions are a way to group a set of programming commands into a useful chunk. This helps to divide your sketch into manageable chunks, making it easier to use.

For example, let's write a sketch that makes the Arduino blink rapidly 10 times when it first starts and then blink steadily once each second thereafter.

Read through the following listing, and then I'll explain what is going on.

```
// sketch 01_02_blink_fast_slow
int ledPin = 13;

void setup()
{
  pinMode(ledPin, OUTPUT);
  flash(10, 100);
}

void loop()
{
  flash(1, 500);
}

void flash(int n, int delayPeriod)
{
  for (int i = 0; i < n; i++)
  {
    digitalWrite(ledPin, HIGH);
    delay(delayPeriod);
    digitalWrite(ledPin, LOW);
    delay(delayPeriod);
  }
}
```

The **setup** function now contains a line that says **flash(10, 100);**. This means flash **10** times with a **delayPeriod** of **100** milliseconds. The **flash** command is not a built-in Arduino command; you are going to create this quite useful function yourself.

The definition of the function is at the end of the sketch. The first line of the function definition is

```
void flash(int n, int delayPeriod)
```

This tells the Arduino that you are defining your own function called **flash** and that it takes two parameters, both of which are **int**s. The first is **n**, which is the number of times to flash the LED, and the second is **delay-Period**, which is the delay to use between turning the LED on or off.

These two parameter variables can only be used inside the function. So, **n** is used in the **for** command to determine how many times to repeat the loop, and **delayPeriod** is used inside the **delay** commands.

The sketch's **loop** function also uses the previous **flash** function, but with a longer **delayPeriod**, and it only makes the LED flash once. Because it is inside **loop**, it will just keep flashing anyway.

## Digital Inputs

To get the most out of this section, you need to find a short length of wire or even a metal paperclip that has been straightened.

Load the following sketch and run it:

```
// sketch 01_03_paperclip
int ledPin = 13;
int switchPin = 7;

void setup()
{
  pinMode(ledPin, OUTPUT);
  pinMode(switchPin, INPUT_PULLUP);
}

void loop()
{
  if (digitalRead(switchPin) == LOW)
  {
    flash(100);
  }
  else
  {
    flash(500);
  }
}

void flash(int delayPeriod)
{
  digitalWrite(ledPin, HIGH);
  delay(delayPeriod);
  digitalWrite(ledPin, LOW);
  delay(delayPeriod);
}
```

Use your wire or paperclip to connect the GND pin to digital pin 7, as shown in Figure 1-13. You can do this with your Arduino plugged in, but

**Figure 1-13**  *Using a digital input*

only after you have uploaded the sketch. The reason is that if on some previous sketch pin 7 had been set to an output, then connecting it to the GND would damage the pin. Since the sketch sets pin 7 to be an input, this is safe.

This is what should happen: when the paperclip is connected, the LED will flash quickly, and when it is not connected, it will flash slowly.

Let's dissect the sketch and see how it works.

First, we have a new variable called **switchPin**. This variable is assigned to pin 7. So the paperclip is acting like a switch. In the **setup** function, we specify that this pin will be an input using the **pinMode** command. The second argument to **pinMode** is not simply **INPUT** but actually **INPUT_PULLUP**. This tells the Arduino that, by default, the input is to be **HIGH**, unless it is pulled **LOW** by connecting it to GND (with the paperclip).

In the **loop** function, we use the **digitalRead** command to test the value at the input pin. If it is **LOW** (the paperclip is in place), then it calls a function called **flash** with a parameter of **100** (the **delayPeriod**). This makes the LED blink fast.

If, on the other hand, the input is **HIGH**, then the commands in the **else** part of the **if** statement are run. This calls the same **flash** function but with a much longer delay, making the LED blink slowly.

The **flash** function is a simplified version of the **flash** function that you used in the previous sketch, and it just blinks once with the period specified.

Sometimes you will connect digital outputs from a module that does not act as a switch, but actually produces an output that is either HIGH or LOW. In this case, you can use **INPUT** rather than **INPUT_PULLUP** in the **pinMode** function.

## Digital Outputs

There is not really much new to say about digital outputs from a programming point of view, as you have already used them with the built-in LED on pin 13.

The essence of a digital output is that in your **setup** function you define them as being an output using this command:

```
pinMode(outputPin, OUTPUT);
```

When you want to set the output **HIGH** or **LOW**, you use the **digitalWrite** command:

```
digitalWrite(outputPin, HIGH);
```

## The Serial Monitor

Because your Arduino is connected to your computer by USB, you can send messages between the two using a feature of the Arduino IDE called the *Serial Monitor*.

To illustrate, let's modify the sketch 01_03 so that, instead of changing the LED blink rate when digital input 7 is LOW, it sends a message.

Load this sketch:

```
// sketch 01_04_serial
int switchPin = 7;

void setup()
{
  pinMode(switchPin, INPUT_PULLUP);
  Serial.begin(9600);
}

void loop()
{
  if (digitalRead(switchPin) == LOW)
  {
    Serial.println("Paperclip connected");
  }
  else
  {
    Serial.println("Paperclip NOT connected");
  }
  delay(1000);
}
```

Now open the Serial Monitor on the Arduino IDE by clicking the icon that looks like a magnifying glass on the toolbar. You should immediately start to see some messages appear, once each second (Figure 1-14).

Disconnect one end of the paperclip, and you should see the message change.

Because you are no longer using the built-in LED, you do not need the **ledPin** variable any more. Instead, you need to use the **Serial.begin** command to start serial communications. The parameter is the baud rate. In Chapter 13, you will find out much more about serial communications.

To write messages to the Serial Monitor, all you need to do is use the **Serial.println** command.

In this example, the Arduino is sending messages to the Serial Monitor.

**Figure 1-14**   *The Serial Monitor*

# Arrays and Strings

*Arrays* are a way of containing a list of values. The variables you have met so far have only contained a single value, usually an **int**. By contrast, an array contains a list of values, and you can access any one of those values by its position in the list.

C, in common with most programming languages, begins its index positions at 0 rather than 1. This means that the first element is actually element zero.

You have already met one kind of array in the last section when you learned about the Serial Monitor. Messages like **"Paperclip NOT connected"** are called *character arrays* because they are essentially collections of characters.

For example, let's teach Arduino to talk gibberish over the Serial Monitor.

The following sketch has an array of character arrays and will pick one at random and display it on the Serial Monitor after a random amount of time. This sketch has the added advantage of showing you how to produce random numbers with an Arduino.

```
// sketch 01_05_gibberish

char* messages[] = {
                    "My name is Arduino",
                    "Buy books by Simon Monk",
                    "Make something cool with me",
                    "Raspberry Pis are fruity"};

void setup()
{
  Serial.begin(9600);
}

void loop()
{
  int delayPeriod = random(2000, 8000);
  delay(delayPeriod);
  int messageIndex = random(4);
  Serial.println(messages[messageIndex]);
}
```

Each of the messages, or *strings* as collections of characters are often called, has a data type of **char***. The * is a pointer to something. We'll get to the advanced topic of pointers in Chapter 6. The [] on the end of the variable declaration indicates that the variable is an array of **char*** rather than just a single **char*** on its own.

Inside the **loop** function, the value **delayPeriod** is assigned a random value between **2000** and **7999** (the second argument to "random" is exclusive). A pause of this length is then set using the **delay** function.

The **messageIndex** variable is also assigned a random value using **random**, but this time **random** is only given one parameter, in which case a random number between 0 and 3 is generated as the index for the message to be displayed.

Finally, the message at that position is sent to the Serial Monitor. Try out the sketch, remembering to open the Serial Monitor.

# Analog Inputs

The Arduino pins labeled A0 to A5 can measure the voltage applied to them. The voltage must be between 0 and 5V. The built-in Arduino function that does this is **analogRead**, and it returns a value between 0 and 1023: 0 at 0V and 1023 at 5V. So to convert that number into a value between 0 and 5, you have to divide 1023/5 = 204.6.

To measure voltage, **int** is not the ideal data type as it only represents whole numbers and it would be good to see the fractional voltage, for which you need to use the **float** data type.

Load this sketch onto your Arduino and then attach the paperclip between A0 and 3.3V (Figure 1-15).

**Figure 1-15** *Connecting 3.3V to A0*

```
// sketch 01_06_analog
int analogPin = A0;

void setup()
{
  Serial.begin(9600);
}

void loop()
{
  int rawReading = analogRead(analogPin);
  float volts = rawReading / 204.6;
  Serial.println(volts);
  delay(1000);
}
```

Open the Serial Monitor, and a stream of numbers should appear (Figure 1-16). These should be close to 3.3.

*CAUTION*    *Do not connect any of the supply voltages together (5V, 3.3V, or GND). Creating such a short circuit would probably damage your Arduino and could even damage your computer.*

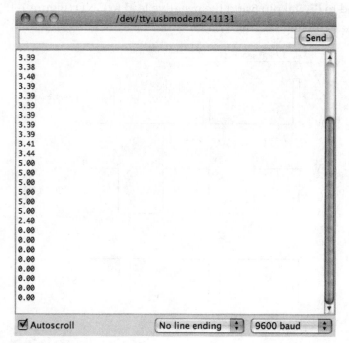

**Figure 1-16**    *Voltage readings*

If you now keep one end of the paperclip in A0 but move the other end of the paperclip to 5V, the readings will change to around 5V. Moving the same end to GND gives you a reading of 0V.

## Analog Outputs

The Arduino Uno does not produce true analog outputs (for that you need an Arduino Due), but it does have a number of outputs that are capable of producing a pulse-width modulation (PWM) output. This approximates to an analog output by controlling the length of a stream of pulses, as you can see in Figure 1-17.

The longer the pulse is high, the higher the average voltage of the signal. Since there are about 600 pulses per second and most things that you would connect to a PWM output are quite slow to react, the effect is of the voltage changing.

On an Arduino Uno, the pins marked with a little ~ (pins 3, 5, 6, 9, 10, and 11) can be used as analog outputs.

If you have a voltmeter, set it to its 0..20V DC range and attach the positive lead to digital pin 6 and the negative lead to GND (Figure 1-18). Then load the following sketch:

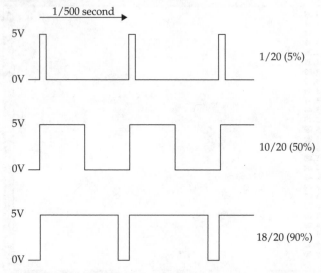

**Figure 1-17** *Pulse-width modulation*

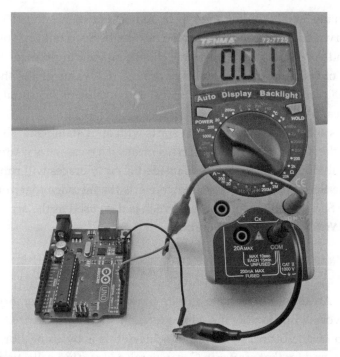

**Figure 1-18**  *Measuring the output voltage*

```
// sketch 01_07_pwm
int pwmPin = 6;

void setup()
{
  pinMode(pwmPin, OUTPUT);
  Serial.begin(9600);
}

void loop()
{
  if (Serial.available())
  {
    int dutyCycle = Serial.parseInt();
    analogWrite(pwmPin, dutyCycle);
  }
}
```

Open the Serial Monitor and type a number between 0 and 255 into the text entry field at the top of the screen next to the Send button. Then press Send and you should see the voltage change on your multimeter. Sending a value of 0 should give a voltage of around 0. A value of 127 should be about halfway between 0 and 5V (2.5V) and a value of 255 should give a value near 5V.

In this sketch, the **loop** function starts with an **if** statement. The condition for the **if** is **Serial.available()**. This means if a message is waiting from the Serial Monitor, the commands inside the curly braces will run. In this case, the **Serial.parseInt** command converts the message that you typed into the Serial Monitor into an **int**, which is then used as the argument to **analogWrite** to set the PWM output.

# Using Libraries

Because Arduino boards have a quite limited amount of memory, you'll find it worthwhile to only include code that will actually be used in what ends up on the board. One way to achieve this is by using libraries. In Arduino, and for that matter in C in general, a *library* is a collection of useful functions.

So, for example, the Arduino IDE includes a library for using an LCD display. This uses about 1.5kB of program memory. There is no point in this library being included unless you are using it, so such libraries are "included" when needed.

You accomplish this using the **#include** directive at the beginning of your sketch. You can add an **include** statement for any libraries that the Arduino IDE has installed using the Sketch | Import Library... menu option.

The Arduino IDE comes with a large selection of "official" libraries, including:

- **EEPROM**   For storing data in EEPROM memory
- **Ethernet**   For network programming
- **Firmata**   The serial communications standard for Arduino to computer

- **LiquidCrystal**   For alphanumeric LCD displays
- **SD**   For reading and writing SD flash memory cards
- **Servo**   For controlling servo motors
- **SPI**   The Arduino to peripheral communication bus
- **Software Serial**   For serial communication using nonserial pins
- **Stepper**   For controlling stepper motors
- **WiFi**   For WiFi network access
- **Wire**   For I2C communication with peripherals

Some libraries are specific to a type of Arduino board:

- **Keyboard**   USB keyboard emulation (Leonardo, Due, and Micro)
- **Mouse**   USB mouse emulation (Leonardo, Due, and Micro)
- **Audio**   Audio playing utilities (Due only)
- **Scheduler**   For managing multiple execution threads (Due only)
- **USBHost**   USB peripherals (Due only)

Finally, there are a huge number of libraries that other Arduino users have written that can be downloaded from the Internet. Some of the more popular ones are

- **OneWire**   For reading data from Dallas Semiconductor's range of digital devices using the 1-wire bus interface
- **Xbee**   For Wireless serial communication
- **GFX**   A graphics library for many different types of display from Adafruit
- **Capacitive Sensing**   For proximity detection
- **FFT**   Frequency analysis library

New libraries appear all the time and you may find them on the official Arduino site (http://arduino.cc/en/Reference/Libraries) or you may find them with an Internet search.

If you want to use one of these last categories of libraries, then you need to install it by downloading the library and then saving it to the Libraries folder within your Arduino folder (in your Documents folder). Note that if there is no Libraries folder, you will need to create it the first time that you add a library.

For the Arduino IDE to become aware of a library that you have installed, you need to exit and restart the IDE.

# Arduino Data Types

A variable of type **int** in Arduino C uses 2 bytes of data. Unless a sketch becomes very memory hungry, then **int**s tend to be used for almost everything, even for Boolean values and small integers that could easily be represented in a single byte value.

Table 1-1 contains a full list of the data types available.

| Type | Memory (bytes) | Range | Notes |
|---|---|---|---|
| boolean | 1 | true or false (0 or 1) | Used to represent logical values. |
| char | 1 | –128 to +128 | Used to represent an ASCII character code; for example, A is represented as 65. Negative numbers are not normally used. |
| byte | 1 | 0 to 255 | Often used for communicating serial data, as a single unit of data. See Chapter 9. |
| int | 2 | –32768 to +32767 | These are signed 16 bit values. |
| unsigned int | 2 | 0 to 65536 | Used for extra precision when negative numbers are not needed. Use with caution as arithmetic with **int**s may cause unexpected results. |
| long | 4 | 2,147,483,648 to 2,147,483,647 | Needed only for representing very big numbers. |
| unsigned long | 4 | 0 to 4,294,967,295 | *See* unsigned int. |
| float | 4 | –3.4028235E+38 to + 3.4028235E+38 | Used to represent floating point numbers. |
| double | 4 | as float | Normally, this would be 8 bytes and higher precision than **float** with a greater range. However, on Arduino **double** is the same as **float**. |

**Table 1-1** *Data Types in Arduino C*

# Arduino Commands

A large number of commands are available in the Arduino library, and a selection of the most commonly used commands is listed, along with examples, in Table 1-2.

| Command | Example | Description |
|---|---|---|
| **Digital I/O** | | |
| pinMode | **pinMode(8, OUTPUT);** | Sets pin 8 to be an output. The alternative is to set it to be **INPUT** or **INPUT_PULLUP**. |
| digitalWrite | **digitalWrite(8, HIGH);** | Sets pin 8 high. To set it low, use the constant **LOW** instead of **HIGH**. |
| digitalRead | **int i;**<br>**i = digitalRead(8);** | Sets the value of **i** to **HIGH** or **LOW**, depending on the voltage at the pin specified (in this case, pin 8). |
| pulseIn | **i = pulseIn(8, HIGH)** | Returns the duration in microseconds of the next **HIGH** pulse on pin 8. |
| tone | **tone(8, 440, 1000);** | Makes pin 8 oscillate at 440 Hz for 1000 milliseconds. |
| noTone | **noTone();** | Cuts short the playing of any tone that was in progress. |
| **Analog I/O** | | |
| analogRead | **int r;**<br>**r = analogRead(0);** | Assigns a value to **r** of between 0 and 1023: 0 for 0V, 1023 if pin0 is 5V (3.3V for a 3V board). |
| analogWrite | **analogWrite(9, 127);** | Outputs a PWM signal. The duty cycle is a number between 0 and 255, 255 being 100%. This must be used by one of the pins marked as PWM on the Arduino board (3, 5, 6, 9, 10, and 11). |

**Table 1-2**   *Arduino Library Functions* (continued)

| Command | Example | Description |
|---|---|---|
| *Time Commands* | | |
| millis | unsigned long l;<br>l = millis(); | The variable type **long** in Arduino is represented in 32 bits. The value returned by **millis()** is the number of milliseconds since the last reset. The number wraps around after approximately 50 days. |
| micros | long l;<br>l = micros(); | See **millis**, except this is microseconds since the last reset. It wraps after approximately 70 minutes. |
| delay | delay(1000); | Delays for 1000 milliseconds or 1 second. |
| delayMicroseconds | delayMicroseconds(100000); | Delays for 100,000 microseconds.<br>Note the minimum delay is 3 microseconds; the max is around 16 milliseconds. |
| *Interrupts (see Chapter 3)* | | |
| attachInterrupt | attachInterrupt(1, myFunction, RISING); | Associates the function **myFunction** with a rising transition on interrupt 1 (D3 on an Uno). |
| detachInterrupt | detachInterrupt(1); | Disables any interrupt on interrupt 1. |

**Table 1-2**  *Arduino Library Functions*

For a full reference to all the Arduino commands, see the official Arduino documentation at http://arduino.cc.

# Summary

By necessity, this chapter has been a very condensed introduction to the world of Arduino. If you require more information about the basics, then there are many online resources, including free Arduino tutorials at http://www.learn.adafruit.com.

In the next chapter, we will dig under the surface of Arduino and see just how it works and what is going on inside the nice, easy-to-use Arduino environment.

# 2
# Under the Hood

**The nice** thing about the Arduino is that a lot of the time, you really do not need to know what goes on behind the scenes when you upload a sketch. However, as you get more into Arduino and want to push the envelope of what it can do, you need to find out a bit more about what's going on behind the scenes.

## A Brief History of Arduino

The first Arduino board was developed back in 2005 in Italy at the Interaction Design Institute at Ivrea near Turin. The intention was to design a low-cost and easy-to-use tool for design students to build interactive systems. The software behind Arduino, which is so much a part of Arduino's success, is a fork of an open source framework called Wiring. Wiring was also created by a student at the Institute.

The Arduino fork of Wiring is still very close to Wiring, and the Arduino IDE is written in Wiring's big brother that runs on PCs, Macs, and so on, and is called Processing. Processing is well worth a look if you have a project where your Arduino needs to talk to a PC over USB or Bluetooth.

The Arduino hardware has evolved over the years, but the current Arduino Uno and Leonardo boards retain the same basic shape and sockets as the original.

## Anatomy of an Arduino

Figure 2-1 shows the anatomy of an Arduino Uno. The Leonardo is similar but has the USB interface integrated into the main microcontroller chip. The Due is also similar, but the processor is powered by 3.3V, not 5V.

In many ways, the Arduino is really little more than a microcontroller chip with some supporting components. In fact, it is perfectly possible to build an Arduino on breadboard using the processor chip and a few extra components or to create a PCB for a design that started out using an Arduino as a prototype. The Arduino boards make things easy, but ultimately any Arduino design can be converted into something that just uses the microcontroller chip and the few components that it really needs. For example, if the design is only for programming purposes, you may not need a USB interface, as you could program the chip on an Arduino and then transplant the programmed chip into an IC socket on a PCB or to breadboard.

Later, we'll look at how Arduinos can be programmed directly using the ICSP (In Circuit Serial Programming) interface.

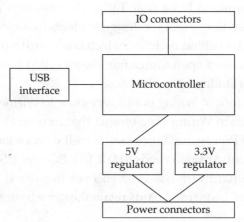

**Figure 2-1** *The anatomy of an Arduino Uno*

# AVR Processors

The Arduino family of boards all use microcontrollers made by Atmel. They all have similar hardware design principals and, with the exception of the microcontroller used in the Due (SAM3X8E ARM Cortex-M3 CPU), they have similar designs.

## ATmega328

The ATmega328 is the microcontroller used in the Arduino Uno and its predecessor the Duemilanove. In fact, the ATmega168 that was used in the first Arduino boards is basically an ATmega328 but with half of each type of memory.

Figure 2-2 shows the internals of an ATmega328, taken from its datasheet. The full datasheet is available from www.atmel.com/Images/doc8161.pdf and is worth browsing through to learn more about the inner workings of this device.

The central processing unit (CPU) is where all the action takes place. The CPU reads instructions (compiled sketch code) from the flash memory one instruction at a time. This process is different from a conventional computer where programs are stored on disk and loaded into random access memory (RAM) before they can be run. Variables that you use in your programs are stored separately in the static RAM (SRAM). Unlike the flash memory containing the program code, the RAM is volatile and loses its contents when you turn off the power.

To allow the nonvolatile storage of data that remains even after the device is powered off, a third type of memory called *Electrically Erasable Programmable Read Only Memory (EEPROM)* is used.

Another area of interest is the Watchdog Timer and Power Supervision unit. These give the microcontroller the capability to do a number of things that are normally hidden by the simplified Arduino layer, including clever tricks like putting the chip to sleep and then setting a timer to wake it up periodically. This trick can be very useful in low current applications, and you can read more on this in Chapter 5.

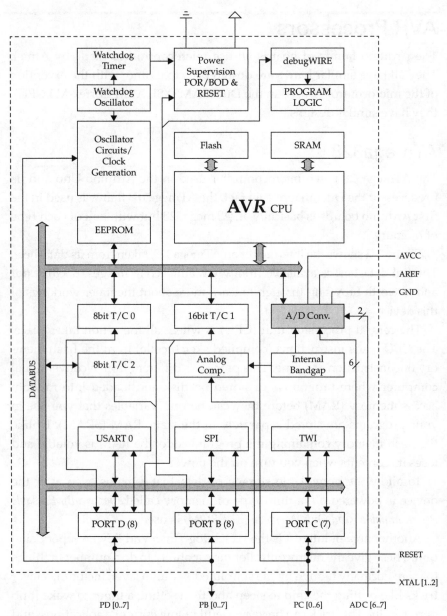

**Figure 2-2** *The ATmega328*

The remainder of the block diagram is concerned with the analog-to-digital conversion, the input/output ports, and the three types of serial interfaces supported by the chip: UART - Serial, SPI, and TWI (I2C).

## ATmega32u4

The ATmega32u4 is used in the Arduino Leonardo and also in the LilyPad USB and the Arduinos Micro and Nano. This processor is similar to the ATmega328, but it is a more modern chip with a few enhancements over the ATmega328:

- A built-in USB interface, so there's no need for extra USB hardware.

- More of the pins are PWM capable.

- There are two serial ports.

- Dedicated pins for I2C (these pins are shared with the analog pins on the Arduino).

- There is 0.5kB more SRAM.

The version used in the Leonardo is in a surface-mount package, which means it is soldered directly to the Arduino board, whereas the ATmega328 is in a DIL package fitted into an IC socket for the Arduino Uno.

## ATmega2560

The ATmega2560 is used in the Arduino Mega 2560 and the Arduino Mega ADK. It is no faster than the other ATmega chips, but it does have far more of every type of memory (256k flash, 8k SRAM, and 4k of EEPROM) and many more I/O pins.

## AT91SAM3X8E

This is the chip at the heart of the Arduino Due. It is much faster than the ATmega chips I have discussed so far, being clocked at 84 MHz, rather than the normal 16 MHz of the ATmegas. It has 512k of flash and 96KB of SRAM. The microcontroller does not have any EEPROM. Instead, to save persistent data, you need to provide your own additional hardware, either in the form of an SD card holder and SD card or flash or EEPROM storage

ICs. The chip itself has many advanced features including two analog outputs that make it ideal for sound generation.

# Arduino and Wiring

The Wiring library gives Arduino its easy-to-use functions for controlling the hardware pins; however, the main structural part of the language is all provided by C.

Until recently, if you looked in your Arduino installation directory, you could still find a file called **WProgram.h** (Wiring Program). This file has now been replaced by a similar file called **Arduino.**h that indicates the gradual drift of the Arduino fork away from the original Wiring Project.

If you go to your Arduino installation folder, you'll find a folder called "hardware," and within that, a folder called "arduino," and within that, a folder called "cores." Note that if you are using a Mac, then you can only get to this folder by right-clicking on your Arduino application, selecting View Package Contents from the menu, and then navigating to the Resources/Java/ folder.

Inside the cores folder is another folder called "arduino," and, in there, you will find a whole load of C header files with the file extension **.h** and C++ implementation files with the extension **.cpp** (Figure 2-3).

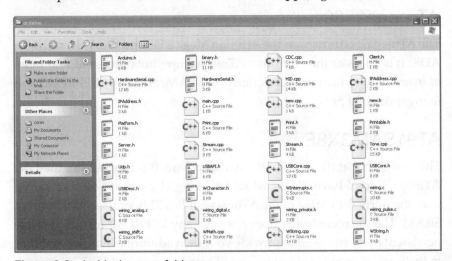

**Figure 2-3**  *Inside the cores folder*

If you open **Arduino.h** in an editor, you'll discover that it consists of many **#include** statements. These pull in definitions from other header files within the cores/arduino folder, so they are included during compilation (converting the sketch into a form suitable for installing into the microcontroller's flash memory).

You'll also find constant definitions like this:

```
#define HIGH 0x1
#define LOW  0x0

#define INPUT 0x0
#define OUTPUT 0x1
#define INPUT_PULLUP 0x2
```

You can think of these as being a bit like variables, so the name **HIGH** is given the value 1. The value is specified as **0x1** rather than just 1 because the values are all specified in *hexadecimal* (number base 16). These are not actually variable definitions; they are called *C precompiler directives*, which means that while your sketch is being turned into something that can be installed into the flash memory on the microcontroller, any instances of the words **HIGH**, **LOW**, and so on, are automatically converted into the appropriate number. This has an advantage over using variables in that no memory has to be reserved for their use.

Because these constants are numbers, you could write something like this in your sketch to set pin 5 to be an **OUTPUT**, but it is better to use the name in case the Arduino developers ever decide to change the constant's value. Using a name also makes the code easier to read.

```
setMode(5, 1);
setMode(5, OUTPUT);
```

Also, within **arduino.h**, you'll find lots of function "signatures" like this:

```
void pinMode(uint8_t, uint8_t);
void digitalWrite(uint8_t, uint8_t);
int digitalRead(uint8_t);
int analogRead(uint8_t);
void analogReference(uint8_t mode);
void analogWrite(uint8_t, int);
```

These warn the compiler about functions whose actual implementations are to be found elsewhere. Take the first one as an example. It specifies the function **pinMode** as taking two arguments (that you know to be pin number and mode) that are specified as having a type of **uint8_t**. The **void** command means the function will not return a value when it is called.

You might be wondering why these parameters' type is specified as **uint8_t** rather than **int**. Normally when defining which pin to use, you specify an **int**. In actual fact, **int** is a universal type when writing sketches. It  means users do not need to worry about a large number of possible types that they might have to use. But in Arduino C, an **int** is actually a 16-bit signed number that can represent a number between −32,768 and 32,767. However, when specifying a pin to use, having negative pin numbers doesn't make sense and you are very unlikely to ever get a 32,767-pin Arduino.

The type **uint_8** is a much more precise convention for defining types because an **int** in C can be anything between 16 and 64 bits, depending on the C implementation. The way to read "**uint_8**" is that the **u** is for unsigned, then you have **int**, and, finally, after the _ you have the number of bits. So **uint_8** is an unsigned 8-bit integer that can represent a number between 0 and 255.

You can use these well-defined types within your sketches, and, indeed, some people do. You have to remember, however, that this makes your code a little less accessible to those who aren't as experienced in Arduino programming.

The reason that using a regular signed 16-bit **int** works, rather than, say, a **unit_8**, is that the compiler automatically performs the conversion for you. Using **int** variables for pin numbers actually wastes memory. However, you have to balance this against the simplicity and readability of the code. Generally, in programming it's better to favor easy-to-read code over minimizing memory usage, unless you know you are doing something complex that is going to push the microcontroller's limits.

It's a bit like having a truck in which you want to deliver some goods to someone. If you have a load of stuff to deliver, then you need to think carefully about how to pack the load so it all fits. If you know that you are only going to use one little corner of the available space, then spending a lot of time minimizing the space it takes is simply unnecessary.

Also within the arduino folder, you'll find a file called **main.cpp**. Open this file; you'll find it pretty interesting.

```
#include <Arduino.h>

int main(void)
{
        init();

#if defined(USBCON)
        USBDevice.attach();
#endif

        setup();

        for (;;) {
                loop();
                if (serialEventRun) serialEventRun();
        }

        return 0;
}
```

If you have done any C, C++, or Java programming before, you are familiar with the concept of a **main** function. This function runs automatically when the program is run. Main is the starting point for the whole program. This is also true of Arduino programs, but it is hidden from the sketch writer, who is instead told to implement two functions—setup and loop—within their sketch.

If you look carefully at **main.cpp**, ignoring the first few lines for now, you can see that it actually calls **setup()** and then has a **for** loop with no conditions, with the **loop** function called inside the loop.

The command **for(;;)** is simply an ugly way of writing **while (true)**. Notice that in addition to running the **loop** function, there is also an **if** command inside the **for** that checks for serial messages and services them if they arise.

Returning to the top of **main.cpp**, you can see that the first line is an **include** command that pulls in all the definitions in the header file **arduino.h** that I mentioned previously.

Next, you see the start of the definition of the **main** function, which begins by invoking an **init()** function. If you look, you can find what this does in the file **wiring.c**; it in turn calls a function **sei**, which enables interrupts.

These lines

```
#if defined(USBCON)
      USBDevice.attach();
#endif
```

are another example of a C preprocessor directive. This code is a bit like an **if** command that you might use in your sketch, but the decision in the **if** is not made when the sketch is actually running on the Arduino. The **#if** is evaluated as the sketch is being compiled. This directive is a great way to switch chunks in and out of the build, depending on whether they are needed for a particular type of board. In this case, if the Arduino supports USB, then include the code for attaching the USB (initialize it); otherwise, there is no point in even compiling the code to do that.

## From Sketch to Arduino

Now that you have a basic understanding of where all the magic code comes from when you write a simple Arduino sketch, let's look at exactly how that code gets into the flash memory of an Arduino board's microcontroller when you click the Upload button in the Arduino IDE.

Figure 2-4 shows what happens when you click the Upload button.

Arduino sketches are held in a text file with the **.ino** extension, in a folder of the same name but without the extension.

What actually happens is that the Arduino IDE controls a number of utility programs that do all the actual work. First, a part of the Arduino IDE that (for want of a better name), I have named the *Arduino IDE preprocessor* assembles the files provided as part of the sketch. Note that normally only one file is in the sketch folder; however, you can place other files in the folder if you wish, but you need to use a separate editor to create them.

If you have other files in the folder, they will be included in this build process. C and C++ files are compiled separately. A line to include **arduino.h** is added to the top of the main sketch file.

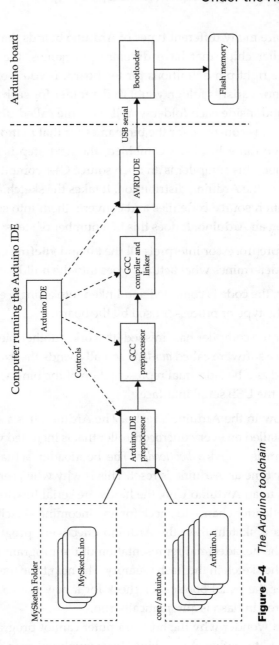

**Figure 2-4**  *The Arduino toolchain*

As there are many different types of Arduino boards that use different microcontroller chips that have different pin names, the Arduino IDE must use the right pin definitions for the board. If you look in the hardware/arduino/variants folder, you'll find a folder for each type of Arduino board, and inside each folder, you'll see a file called **pins_arduino.h**. This file contains constants for the pin names for that platform.

When everything has been combined, the next step is to invoke the GCC compiler. This compiler is an open source C++ compiler that is bundled as part of the Arduino distribution. It takes the sketch, header, and C implementation source code files and converts them into something that can be run on an Arduino. It does this in a number of steps:

1. The preprocessor interprets all the **#if** and **#define** commands and determines what actually goes into the build.

2. Next, the code is compiled and linked into a single executable file for the type of processor used by the board.

3. After the compiler has finished its work, another piece of open source software called *avrdude* actually sends the executable code, saved as a hexadecimal representation of the binary, to the board over the USB serial interface.

We are now in the Arduino's realm. The Arduino has a small resident program installed on every microcontroller that is included with its board. This program is called a *bootloader*. The bootloader actually runs very briefly every time an Arduino is reset. This is why when serial communication starts to an Arduino Uno, the hardware serial link forces a reset to give the bootloader chance to check for any incoming sketches.

If there is a sketch, then the Arduino effectively programs itself by unpacking the hexadecimal representation of the program into binary. It then stores the sketch in the flash memory. The next time that the Arduino restarts, after the usual bootloader check for a new sketch, the program that was stored in flash is automatically run.

You might wonder why the host computer cannot program the microcontroller directly rather than taking this convoluted path. The reason is that programming a microcontroller requires special hardware that uses a different interface to the Arduino board (ever wondered what the little six-pin header was for?). By using a bootloader that can listen on a serial port, you can program the Arduino though USB without having to use special programming hardware.

However, if you do have such a programmer, such as the AVRISPv2, AVRDragon, or the USBtinyISP, then you can program the Arduino directly through such a programmer, bypassing the bootloader entirely. In fact, as you shall see later in this chapter, you can also use a second Arduino as a programmer.

# AVR Studio

Certain hard-bitten electronic engineers can be a bit snotty about Arduino. They might tell you that it doesn't have any technical advantages over using the tools provided by Atmel for programming the whole family of AVR microcontrollers. While technically true, this misses the point of Arduino, which is to demystify the whole process of using a microcontroller and to wrestle it from the control of such experts. This does mean that some of the things us Arduino aficionados do could be considered a bit amateurish, but I say so what!

AVR Studio is the manufacturer's proprietary software for programming the microcontrollers used in Arduinos. You can use it to program the Arduino itself, rather than using the Arduino IDE. If you do, however, you will have to accept the following:

- A Windows-only environment
- Using a hardware programmer rather than USB
- A more complex environment

Perhaps this is the point at which you might want to consider why you might want to do this. Here are some good reasons:

- You want to get rid of the bootloader (it uses 500 bytes on a Uno) because either you are short of flash memory or you want a quicker start after reset.
- You want to target other microcontrollers than those used in standard Arduinos, such as the less expensive and smaller ATtiny family.
- You just want to learn something new.

The Arduino boards all come with a six-pin header that can be used to program the Arduino directly using AVR Studio. In fact, some boards come with two six-pin headers: one for the main processor and one for the USB interface, so be careful to connect to the right one.

Figure 2-5 shows AVR Studio 4 in action.

It is beyond the scope of this book to teach AVR Studio. However, as you can see from Figure 2-5, the Blink sketch does not get any longer, but it certainly looks more complicated! It will also compile into a tiny amount of flash memory compared with its Arduino counterpart.

Figure 2-6 shows an Arduino connected to an AVR Dragon programmer. This programmer is particularly powerful and flexible, and it allows you to debug and single-step through programs actually running on the ATmega chip.

In Chapter 4, we look at the kind of direct port manipulation that is going on in Figure 2-5 as a way to improve I/O performance without having to abandon the Arduino IDE.

**Figure 2-5**   *AVR Studio*

**Figure 2-6**   *An Arduino connected to an AVR Dragon programmer*

# Installing a Bootloader

You might want to install the Arduino bootloader onto an Arduino board for several reasons. You may have damaged the removable ATmega328 on an Arduino Uno and be replacing the chip with a new ATmega328 (bought without the bootloader). Alternatively, you may be moving an Arduino prototype off-board, by taking the ATmega328 off the Arduino board and fitting it to a custom board of your own design.

Whatever the reason, you can add a bootloader to a blank ATmega328, either by using one of the programmers mentioned in the previous section or by using one Arduino to program a second.

## Burning a Bootloader with AVR Studio and a Programmer

The Arduino installation folder contains bootloader hex files that can be flashed onto an ATmega328 using AVR Studio. You will find these files in the hardware/arduino/bootloaders folder. There, you will find hex files for all sorts of different hardware. If want to install a bootloader for an Uno, use the **optiboot_atmega328.hex** file in the optiboot folder (Figure 2-7).

First, a word of warning. If you are going to try this, then be aware that there is a chance you will "brick" your processor chip. These chips have what are called "fuses" that can be set and sometimes cannot be reset. They are designed this way for proprietary reasons, when you want to prevent reprogramming for commercial reasons. Check carefully that the fuses are set correctly for the Arduino board you are programming before you take the plunge, and accept that you may incur a loss. The Arduino forum at www.arduino.cc/forum includes many threads on this topic, along with "gotchas" to avoid.

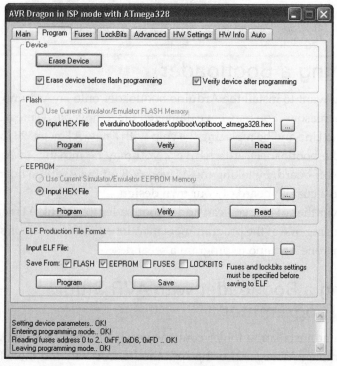

**Figure 2-7**   *Burning an Uno bootloader in AVR Studio 4*

To burn the bootloader using AVR Studio and an AVR Dragon, connect the programmer to the Arduino ISP header pins (see Figure 2-6 earlier in the chapter). Note that an Arduino Uno actually has two sets of ISP header pins; the other sets are for flashing the USB interface.

From the Tools menu, select the Program AVR option and then connect to the ATmega328 on the Arduino. Then in the Flash section, browse to the correct hex file and then click Program.

# Burning a Bootloader with the Arduino IDE and a Second Arduino

Flashing one Arduino with a new bootloader from another Arduino is remarkably easy. It is certainly easier and less risky than using AVR Studio. The Arduino IDE includes an option to do this. Here's all you need to get started:

- Two Arduino Unos
- Six male-to-male jumper leads (or solid core wire)
- One short length of solid core wire
- A 10µF 10V (100µF will also work) capacitor

You first need to make the connections listed in Table 2-1.

You also need to attach a 10µF capacitor between Reset and GND on the target Arduino (the one being programmed). The capacitor will have a longer positive lead, which should go to Reset.

| Arduino Acting as Programmer | Target Arduino |
|---|---|
| GND | GND |
| 5V | 5V |
| 13 | 13 |
| 12 | 12 |
| 11 | 11 |
| 10 | Reset |

**Table 2-1**  *Arduino to Arduino Programming Connections*

**Figure 2-8**  *Arduino to Arduino Flashing*

Figure 2-8 shows the connected Arduinos. The Arduino on the right of Figure 2-8 is the one doing the programming. Notice how solid-core wire is used for the connection between pin 10 on the programming Arduino and Reset on the target Arduino. This is so that both the wire and the positive lead of the capacitor will fit in the Reset socket.

Note that the Arduino doing the programming powers the Arduino being programmed, so only the programming Arduino needs to be connected to your computer by USB.

The Arduino that is going to do the programming needs to have a sketch installed on it. You will find this under File | Examples. The sketch is called **ArduinoISP** and is in the top section of the Examples.

Select the board type and port in the usual way and upload the **ArduinoISP** sketch onto the Arduino doing the programming. Now from the Tools menu, select the Programmer submenu and select the Arduino as ISP option.

Finally, select the Burn Bootloader option from the Tools menu. The process takes a minute or two, during which the Rx and Tx LEDs should flicker on the programming Arduino and the "L" LED flicker on the target Arduino.

When the process finishes, that's it—the microcontroller on the target Arduino has a new bootloader installed.

## Summary

In this chapter, we looked more closely at what exactly the Arduino is and how it works. In particular, I showed you what is hidden by the Arduino environment.

In the next chapter, we look at using interrupts and at how to make the Arduino respond to external timer-triggered events using interrupts.

# 3

# Interrupts and Timers

Interrupts allow microcontrollers to respond to events without having to poll continually to see if anything has changed. In addition to associating interrupts with certain pins you can also use timer-generated interrupts.

## Hardware Interrupts

As an example of how to use interrupts, let's revisit digital inputs. The most common way to detect when something has happened at an input (say a switch has been pressed) is to use some code like this:

```
void loop
{
   if (digitalRead(inputPin) == LOW)
   {
      // do something
   }
}
```

This code means we continually check **inputPin** and the moment it reads **LOW**, we do whatever is specified at the **//do something** comment. This process works well, but what if you have a lot of other things to do inside the loop, too? These other things take time, so you could potentially miss a very quick button press because the processor is busy doing something else. In actual fact, with a switch, missing the button press is unlikely as it remains pressed for what in microcontroller terms is a long time.

But what about shorter pulses from a sensor, which may only be active for a few millionths of a second? For these cases, you can use interrupts to receive such events, setting a function to run whenever these events happen, irrespective of whatever else the microcontroller might be doing. Thus, these are called *hardware interrupts.*

On the Arduino Uno, you can only use two pins as hardware interrupts, which is one reason they are used sparingly. The Leonardo has four interrupt-capable pins; bigger boards like the Mega2560 have many more; and, on the Due, you can attach all the pins to interrupts.

The following shows how hardware interrupts work. To try this example, you need some breadboard, a tactile push switch, a 1 kΩ resistor, and some jumper wires.

Figure 3-1 shows the arrangement. The resistor pulls the interrupt pin (D2) HIGH until the button on the switch is pressed, at which point D2 is grounded and goes LOW.

**Figure 3-1**  *Interrupt test circuit*

Load the following sketch onto your Arduino:

```
// sketch 03_01_interrupts

int ledPin = 13;

void setup()
{
  pinMode(ledPin, OUTPUT);
  attachInterrupt(0, stuffHappened, FALLING);
}

void loop()
{
}

void stuffHappened()
{
  digitalWrite(ledPin, HIGH);
}
```

As well as setting the LED pin to be an output, the **setup** function also contains a line that associates a function with an interrupt. So whenever the interrupt occurs, the function is run. Let's look at this line closely because its arguments are a little confusing:

```
attachInterrupt(0, stuffHappened, FALLING);
```

The first argument **0** is the interrupt number. It would make far more sense if this were a regular Arduino pin number, but it isn't. On an Arduino Uno, interrupt 0 is pin D2 and interrupt 1 is D3. The situation is made even more confusing because on other types of Arduino, these pins are not the same, and on the Arduino Due, you just specify the pin name. When using an Arduino Due, all the pins can be used as interrupts.

I'll come back to this later, but for now let's move on to the second argument. This argument **stuffhappened** is the name of the function to be called when an interrupt occurs. You can see the function defined later in the sketch. Such functions have a special name; they are called *Interrupt Service Routines*, or *ISRs* for short. They cannot have any parameters and should not return anything. This makes sense: Although you can generally call them from other parts of your sketch, no line of code will have called the ISR, so there is no way for them to be given any parameters or to return a value.

The final **attachInterrupt** parameter is a constant, in this case, FALLING. This means the interrupt only results in the ISR being called if D2 goes from HIGH to LOW (in other words, it "falls"), which is what happens when the button is pressed: D2 goes from HIGH to LOW.

You'll notice there is no code in the **loop** function. Normally, the loop function would contain code that would be executed until the interrupt occurred. The ISR itself simply turns the "L" LED on.

When you try the experiment, after the Arduino has reset, the "L" LED should go out. Then as soon as you press the button, the "L" LED should immediately light up and stay lit.

Now change the final argument of **attachInterrupt** to **RISING** and upload the modified sketch. The LED should still remain unlit after the Arduino has finished restarting because the interrupt may be HIGH, but it has always been HIGH; it hasn't, at any point, gone LOW to then "rise" to HIGH.

When you press and hold the button, the LED should stay unlit until you release it. Releasing it triggers the interrupt because D2, which was LOW while the button was pressed, only rises to HIGH when you release it.

If this doesn't seem to work, then the switch is probably bouncing. There isn't a perfect jump from open to closed; rather, the switch will actually turn on and off a few times before settling into the on position. Try it several times, pressing the switch firmly, and you should be able to get a close without a bounce.

The other way to test this is to hold the switch while you press the Reset button on the Arduino. Then when you are ready, release the test button and the "L" LED will light.

## Interrupt Pins

Returning to the thorny issue of how interrupts are named, Table 3-1 shows how the most common Arduino boards map interrupt numbers to physical Arduino pins.

| Board | Interrupt Number | | | | | | Notes |
|---|---|---|---|---|---|---|---|
| | 0 | 1 | 2 | 3 | 4 | 5 | |
| Uno | D2 | D3 | – | – | – | – | |
| Leonardo | D3 | D2 | D0 | D1 | – | – | Yes, really—the interrupt numbers are the opposite of those on the Uno. |
| Mega2560 | D2 | D3 | D21 | D20 | D19 | D18 | |
| Due | | | | | | | Pin numbers are used instead of interrupt numbers |

**Table 3-1**  *Interrupt Pins of Different Arduino Boards*

The pin swap for the first two interrupts on the Uno and Leonardo is an easy trap to fall into. The Due approach of using the Arduino pin name instead of the interrupt number is a much more logical way of doing things.

## Interrupt Modes

The RISING and FALLING modes, which we used in the previous example, are the most handy modes. There are, however, some other interrupt modes. Table 3-2 lists these modes, along with a description.

## Enabling Internal Pull-Up

The hardware setup in the previous example uses a pull-up resistor. Often, the signal that causes the interrupt is from a sensor's digital output, in which case, you do not need a pull-up resistor.

| Mode | Operation | Discussion |
|---|---|---|
| LOW | Triggers interrupt whenever LOW. | This mode sets the ISR to run continuously as long as the pin is low. |
| RISING | Triggers when the pin goes from LOW to HIGH. | |
| FALLING | Triggers when the pin goes from HIGH to LOW. | |
| CHANGE | Triggers whenever the pin changes in either direction. | |
| HIGH | Triggers interrupt whenever HIGH. | This mode is only available on the Due and like LOW is rarely used. |

**Table 3-2**  *Interrupt Modes*

If, however, the sensor is a switch, wired in the same way as the test board shown in Figure 3-1, you can reduce the component count by a resistor if you enable the internal pull-up resistor (about 40 kΩ). To do this, you need to define the interrupt pin explicitly as being an **INPUT_PULLUP** type by adding the bold line, shown here, to the **setup** function:

```
void setup()
{
  pinMode(ledPin, OUTPUT);
  pinMode(2, INPUT_PULLUP);
  attachInterrupt(0, stuffHappened, RISING);
}
```

## Interrupt Service Routines

Sometimes the idea of being able to interrupt what is going on in the **loop** function can seem like an easy way to catch keypresses and so on. But actually there are some fairly strict conditions regarding what you can reliably do within an ISR.

The first thing is that you normally need to keep an ISR as short and fast as possible. If another interrupt occurs while an ISR is running, then the ISR will not itself be interrupted; instead, the interrupt signal is ignored until the ISR has finished. This means that if, for example, you are using the ISR to measure a frequency, you could end up with an incorrect value.

Also, while the ISR is running, nothing happens with the code in the **loop** function until the ISR has finished.

While inside an ISR, interrupts are automatically turned off. This prevents the potential confusion caused by ISRs interrupting each other, but it has some side effects. The **delay** function uses timers and interrupts, so that won't work. The same is true of **millis**. And although **delay** uses **millis** and it will tell you the milliseconds elapsed since reset at the point that the ISR started executing, but it will not change as the ISR runs. However, you can use **delayMicroseconds** because this does not use interrupts.

Serial communication also uses interrupts, so do not use **Serial.print** or try to read from **Serial**. Well, you can try, and it may work, but do not expect it to work reliably all the time.

# Volatile Variables

Because the ISR function is not allowed to take parameters and cannot return a value, you need a way to pass data between the ISR and the rest of the program. You typically do this using global variables, as the next example illustrates:

```
// sketch 03_02_interrupt_flash

int ledPin = 13;
volatile boolean flashFast = false;

void setup()
{
  pinMode(ledPin, OUTPUT);
  attachInterrupt(0, stuffHapenned, FALLING);
}

void loop()
{
  int period = 1000;
  if (flashFast) period = 100;
  digitalWrite(ledPin, HIGH);
  delay(period);
  digitalWrite(ledPin, LOW);
  delay(period);
}

void stuffHapenned()
{
  flashFast = ! flashFast;
}
```

This sketch uses a global variable **flashFast** in the **loop** function to determine the delay period. The ISR then toggles this same variable between **true** and **false**.

Notice that the declaration of the variable **flashFast** includes the word "volatile." You may get away with the sketch working if you do not use **volatile**, but you should use it because if a variable is not declared as being volatile, the C compiler may generate machine code that caches its value in a register to improve performance. If, as is the case here, this caching process could be interrupted, then the variable might not be updated correctly.

## ISR Summary

Keep these points in mind when writing an ISR:

- Keep it fast.
- Pass data between the ISR and the rest of the program using **volatile** variables.
- Don't use **delay**, but you can use **delayMicroseconds**.
- Don't expect serial communications, reading, or writing to be reliable.
- Don't expect the value returned by **millis** to change.

# Enabling and Disabling Interrupts

By default, interrupts are enabled in a sketch and, as I mentioned previously, are automatically disabled when you are inside an ISR. However, you can explicitly turn interrupts on and off from your program code using the functions **interrupts** and **noInterrupts**. Neither function takes any parameters and they turn all interrupts on or off, respectively.

You might want to explicitly turn interrupts on and off if you have an area of code that you do not wish to be disturbed, for example, if you are writing serial data or generating pulses with accurate timing using **delay-Microseconds**.

# Timer Interrupts

As well as interrupts being triggered by external events, you can also trigger ISRs to be called as a result of timed events. This capability can be really useful if you need to do something time-critical.

TimerOne makes it easy to set timed interrupts. You can download the TimerOne library from http://playground.arduino.cc/Code/Timer1.

The following example shows how you can use TimerOne to generate a 1-kHz square wave signal. If you have an oscilloscope or multimeter with a frequency setting, connect it to pin 12 to see the signal (Figure 3-2).

```
// sketch_03_03_1kHz

#include <TimerOne.h>
```

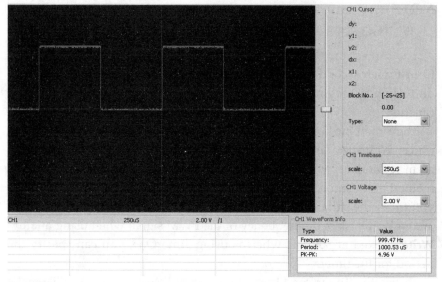

**Figure 3-2**  *A timer-generated square wave*

```
int outputPin = 12;
volatile int output = LOW;

void setup()
{
  pinMode(12, OUTPUT);
  Timer1.initialize(500);
  Timer1.attachInterrupt(toggleOutput);
}

void loop()
{
}

void toggleOutput()
{
  digitalWrite(outputPin, output);
  output = ! output;
}
```

Although you could have written this using **delay**, by using a timer interrupt, you can do other things inside the loop. Also, if you used **delay**, then the frequency would not be as accurate because the actual time to set the output high would not be accounted for in the delay.

*NOTE*  *All the constraints on what you can do in the ISR for external interrupts also apply to timed interrupts.*

You can set the timer interrupt period using this method to anything from 1 microsecond to 8,388,480 microseconds, or about 8.4 seconds. You do this by specifying a period in microseconds in the timer's **initialize** function.

The TimerOne library also allows you to use the timer to generate PWM (Pulse Width Modulation) signals on Arduino pins 9 and 10. This may seem redundant, as you can do that with **analogWrite** anyway, but this method gives you better control of the PWM signal. In particular, it allows you to set the duty cycle between 0 and 1023 rather than the 0 to 255 of **analogWrite**. Also, the frequency of the PWM signal when using **analogWrite** is fixed at 500 Hz, whereas using TimerOne, you can specify the period for the timer.

To use the TimerOne library to generate PWM signals, use **Timer1**'s **pwm** function, as shown in the following code example:

```
// sketch_03_04_pwm
#include <TimerOne.h>

void setup()
{
  pinMode(9, OUTPUT);
  pinMode(10, OUTPUT);
  Timer1.initialize(1000);
  Timer1.pwm(9, 512);
  Timer1.pwm(10, 255);
}

void loop()
{
}
```

In this case, I have set the overall period to 1000 microseconds, resulting in a PWM frequency of 1kHz. Figure 3-3 shows the waveforms generated on pin 10 (top) and pin 9 (bottom).

As an experiment, let's see how far you can push the PWM frequency. Changing the period to 10 results in a PWM frequency of 100 kHz. The waveforms for this are shown in Figure 3-4.

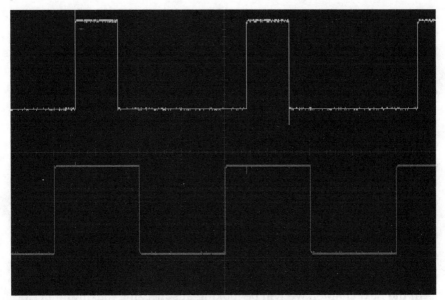

**Figure 3-3**   *Using TimerOne to generate PWM at 1 kHz*

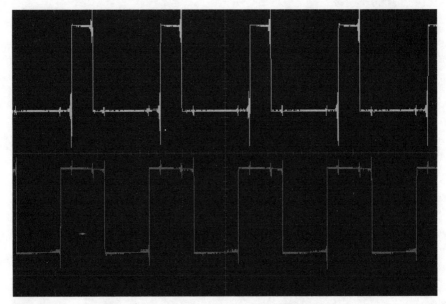

**Figure 3-4**   *Using TimerOne to generate a 100-kHz PWM*

Although there is, as you would expect, a fair amount of transient noise on the signals, you can see that the duty cycles still look pretty close to 25 percent and 50 percent, respectively.

## Summary

Interrupts, although they sometimes seem like the ideal solution to a difficult project, can make the code difficult to debug and are by no means always the best way to handle tasks. Think carefully before you jump into using them. In Chapter 14, we'll explore a different technique for getting around the Arduino's apparent inability to do more than one thing at a time.

We will also return to interrupts in Chapter 5, where we'll look at using them to save power by waking a sleeping Arduino periodically, and in Chapter 13, where we'll use them for accurate timing during digital signal processing.

In the next chapter, we will be looking at squeezing maximum performance out of an Arduino.

# 4

# Making Arduino Faster

This chapter is all about finding out how fast your Arduino is and squeezing it for that extra bit of horse-power when you need it.

## How Fast Is an Arduino?

Before you start worrying about improving the speed of your sketches, let's take a moment to benchmark your Arduino to see just how it compares with other computers, starting with the oft-quoted MHz and GHz.

An Arduino Uno is clocked at 16 MHz. As most instructions (adding or storing a value in a variable) are executed in a single clock cycle, that means the Uno can do 16 million things in one second. Sounds pretty good, doesn't it? The story is not that simple, however, as the C that you write in a sketch expands into quite a few instructions.

Now contrast that with the author's aging Mac laptop that has two processors that are each clocked at 2.5 GHz. My laptop has a clock frequency of over 150 times the frequency of the Arduino. Albeit, the processor takes a few more clock cycles to do each instruction, but as you would expect, it is a lot faster.

Let's try running the following test program on both an Arduino and a slightly modified version on my Mac:

```
// sketch 04_01_benchmark

void setup()
{
```

```
Serial.begin(9600);
Serial.println("Starting Test");
long startTime = millis();

// test code here
long  i = 0;
long j = 0;
for (i = 0; i < 2000000000; i ++)
{
   j = i + i * 10;
   if (j > 10) j = 0;
}
// end of test code
long endTime = millis();

Serial.println(j); // prevent loop being optimized out
Serial.println("Finished Test");
Serial.print("Seconds taken: ");
Serial.println((endTime - startTime) / 10001);
}

void loop()
{

}
```

**NOTE**   *You can find the C counterpart to this code in the download area for code on the book's website.*

Here are the results: on a 2.5-GHz MacBook Pro, the test program took 0.068 seconds to run, whereas on an Arduino Uno, the code took 28 seconds to execute. The Arduino is roughly 400 times slower for this particular task.

# Comparing Arduino Boards

Table 4-1 shows the result of running this test on a few different Arduino boards.

As you can see, the results for most of the boards are consistent, however, the Due results are impressive—more than ten times faster than the other boards.

| Board | Time to Complete Task (seconds) |
|---|---|
| Uno | 28 |
| Leonardo | 29 |
| Arduino Mini Pro | 28 |
| Mega 2560 | 28 |
| Due | 2 |

**Table 4-1**  *Arduino Performance Test Results*

# Speeding Up Arithmetic

As an exercise let's change the benchmark code that we just used and do the arithmetic with floats rather than longs. Both are 32-bit numbers, so you might expect the time to complete the task to be similar. An Arduino Uno is used in the following test.

```
// sketch 04_02_benchmark_float

void setup()
{
  Serial.begin(9600);
  while (! Serial) {};
  Serial.println("Starting Test");
  long startTime = millis();

  // test code here
  long  i = 0;
  float j = 0.0;
  for (i = 0; i < 20000000; i ++)
  {
    j = i + i * 10.0;
    if (j > 10) j = 0.0;
  }
  // end of test code
  long endTime = millis();

  Serial.println(j); // prevent loop being optimized out
  Serial.println("Finished Test");
  Serial.print("Seconds taken: ");
  Serial.println((endTime - startTime) / 1000l);
}

void loop()
{

}
```

Unfortunately, the task takes a lot longer using floats. This example takes the Arduino some 467 seconds instead of 28. So, by changing to floats, my code became about 16 times slower than when I used doubles. To be fair, some of that performance cost was probably also due to converting between float and integer types, which is also quite costly in terms of time.

## Do You Really Need to Use a Float?

A common misconception is that if you are measuring something like temperature, then you need to store it in a float because it will often be a number like 23.5. In fact, you may sometimes wish to display the temperature as a float, but you do not need to store it as a float in your sketch.

An analog input results in an **int** being read, in fact, only 12 bits of an **int**, which is a number between 0 and 1023. You can put those 12 bits into a 32-bit float if you like, but you will not be making the data any more accurate or precise.

This sensor reading could, for example, correspond to a temperature in degrees Celsius (C). One commonly used sensor (the TMP36) has an output voltage proportional to the temperature. The flowing calculation can often be found in sketches to convert an analog reading between 0 and 1023 into a temperature in degrees C.

```
int raw = analogRead(sensePin);
float volts = raw / 205.0;
float tempC = 100.0 * volts - 50;
```

But you actually only need to represent that number in floating point form when you display it. Other things you need to do with the temperature, for example, comparing it or averaging several temperature readings, will be much faster if the arithmetic is done in the temperature's raw **int** state.

## Lookup vs. Calculate

As you have seen, it's best to avoid floats. But if you want to make a sine wave using an analog output, then, as the word *sine* suggests, you need to use the math **sin** function to "draw" the waveform on the analog output. To plot a sine wave on the analog output, you step an angle through $2\pi$ radians, and the value that you send to the analog output is the sin of that angle. Well, actually it's a bit more complicated because you need to center the waveform about an analog output of half the maximum.

The following code generates a sine wave in 64 steps per cycle on an Arduino Due's DAC0 output. Note that only an Arduino with true analog output like the Due works for this experiment.

```
// sketch_04_03_sin

void setup()
{

}
float angle = 0.0;
float angleStep = PI / 32.0;

void loop()
{
   int x = (int)(sin(angle) * 127) + 127;
   analogWrite(DAC0, x);
   angle += angleStep;
   if (angle > 2 * PI)
   {
     angle = 0.0;
   }
}
```

Measuring the signal on the output does, indeed, produce a nice sine wave at a frequency of just 310 Hz. The Arduino Due's processor is clocked at 80 MHz, so you might have expected to generate a faster signal. The problem here is that you are repeating the same calculations again and again. Since they are the same every time, why don't we just generate the values once and store them in an array?

The following code also generates a sine wave with 64 steps, but uses a lookup table of values that are ready to be written straight to the DAC.

```
byte sin64[] = {127, 139, 151, 163, 175, 186, 197,
207, 216, 225, 232, 239, 244, 248, 251, 253, 254,
253, 251, 248, 244, 239, 232, 225, 216, 207, 197, 186,
175, 163, 151, 139, 126, 114, 102, 90, 78, 67, 56, 46,
37, 28, 21, 14, 9, 5, 2, 0, 0, 0, 2, 5, 9, 14, 21, 28,
37, 46, 56, 67, 78, 90, 102, 114, 126};
```

```
void setup()
{
}

void loop()
{
  for (byte i = 0; i < 64; i++)
  {
    analogWrite(DAC0, sin64[i]);
  }
}
```

The waveform generated by this code looks just like the one from the previous example, except that it has a frequency of 4.38 kHz, which is about 14 times faster.

You can calculate the table of sin values in several ways. You can generate the numbers using nothing more complex than a spreadsheet formula, or you can write a sketch that writes the numbers to the Serial Monitor, where they can be pasted into the replacement sketch. Here is an example that modifies **sketch_04_03_sin** to print the values once to the Serial Monitor.

```
// sketch_04_05_sin_print

float angle = 0.0;
float angleStep = PI / 32.0;

void setup()
{
  Serial.begin(9600);
  Serial.print("byte sin64[] = {");
  while (angle < 2 * PI)
  {
    int x = (int)(sin(angle) * 127) + 127;
    Serial.print(x);
    angle += angleStep;
    if (angle < 2 * PI)
    {
      Serial.print(", ");
    }
  }
  Serial.println("};");
}

void loop()
{
}
```

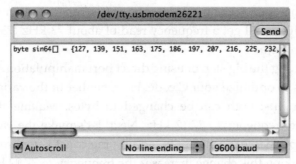

**Figure 4-1**   *Using a sketch to generate code*

Opening the Serial Monitor reveals the code that has been generated (Figure 4-1).

# Fast I/O

In this section, we'll look at how you can improve the speed when turning digital output pins on and off. We'll improve a basic maximum frequency from 73 kHz up to nearly 4 MHz.

## Basic Code Optimization

Let's start with the basic code to turn a digital I/O pin on and off using **digitalWrite**:

```
// sketch_04_05_square

int pin = 10;
int state = 0;

void setup()
{
  pinMode(pin, OUTPUT);
}

void loop()
{
  digitalWrite(pin, state);
  state = ! state;
}
```

If you run this code with an oscilloscope or frequency counter attached to digital pin 10, you'll get a frequency read of about 73 kHz (73.26 kHz on my oscilloscope).

Before taking the big step of using direct port manipulation, you can do a few things to optimize your C code. First, neither of the variables needs to be 16-bit **int**s; both can be changed to bytes. Making this change increases the frequency to 77.17 kHz. Next, let's make the variable containing the pin name a constant by adding the **const** keyword before the variable. Making this change increases the frequency to 77.92 kHz.

In Chapter 2, you learned that the **loop** function is more than just a **while** loop as it also checks for serial communication. Therefore, the next step in improving the performance is to abandon the main **loop** function and move the code into setup. The code containing all these modifications is shown here:

```
// sketch_04_08_no_loop

const byte outPin = 10;
byte state = 0;

void setup()
{
  pinMode(outPin, OUTPUT);
  while (true)
  {
    digitalWrite(outPin, state);
    state = ! state;
  }
}

void loop()
{
}
```

This further improves performance, giving us a new maximum frequency of 86.39 kHz.

Table 4-2 summarizes the improvements that you can make to the basic Arduino code, before taking the final step of abandoning **digitalWrite** for something faster.

|  | Sketch | Frequency |
|---|---|---|
| Original code | 04_05 | 73.26 kHz |
| Bytes instead of **ints** | 04_06 | 77.17 kHz |
| Constant for pin variable | 04_07 | 77.92 kHz |
| Moving loop into setup | 04_08 | 86.39 kHz |

**Table 4-2**   *Speeding Up the Arduino Code*

# Bytes and Bits

Before you can manipulate the I/O ports directly, you need to understand a little about binary, bits, bytes, and **ints**.

Figure 4-2 shows the relationship between bits and bytes.

A *bit* (which is short for *binary digit*) can have one of just two values. It can either be 0 or 1. A *byte* is a collection of 8 bits. Because each of those bits can be either a 1 or a 0, you can actually make 256 different combinations. A byte can be used to represent any number between 0 and 255.

Each of those bits can also be used to indicate if something is on or off. So if you want to turn a particular pin on and off, you need to set a bit to 1 to make a particular output HIGH.

# ATmega328 Ports

Figure 4-3 shows the ports on an ATmega328 and how they relate to the digital pins on an Arduino Uno.

It is no accident that each port has 8 bits (a byte), although ports B and C only use 6 of the bits. Each port is controlled by three *registers*. A register can be thought of as a special variable that you can assign a value to or read the value of. The registers for port D are shown in Figure 4-4.

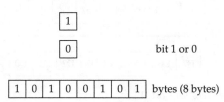

**Figure 4-2**   *Bits and bytes*

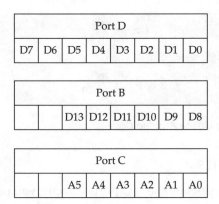

**Figure 4-3**    *ATmega328 ports*

The *data direction register D (DDRD)* has 8 bits, each of which determines whether the corresponding pin on the microcontroller is to be an input or an output. If that bit is set to a 1, the pin is an output; otherwise, it is an input. The Arduino **pinMode** function uses this.

The *PORTD* register is used to set outputs, so a **digitalWrite** sets the appropriate bit for a pin to be a 1 or a 0 (HIGH or LOW).

The final register is called *port input D (PIND)*. By reading this register, you can determine which bits of the port are set HIGH and which are set LOW.

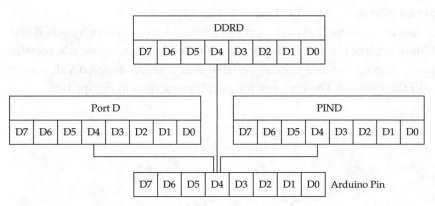

**Figure 4-4**    *The registers for port D*

Each of the three ports has its own three ports, so for port B, they are called DDRB, PORTB, and PINB, and for Port C, they are DDRC, PORTC, and PINC.

## Very Fast Digital Output

The following code uses the ports directly, rather than **pinMode** and **digitalWrite**:

```
// sketch_04_09_square_ports

byte state = 0;

void setup()
{
  DDRB = B00000100;
  while (true)
  {
    PORTB = B00000100;
    PORTB = B00000000;
  }
}

void loop()
{
}
```

Here, we're switching pin D10, which belongs to port B, so first we set the third bit from the left (D10) to be a 1. Note the use of a binary constant **B00000100**. In the main loop, all you have to do is first set the same bit to 1 and then set it to 0 again. You do this simply by assigning a value to PORTB, as if it was a variable.

When this code is run, it generates a frequency of 3.97 MHz (Figure 4-5)—nearly 4 million pulses per second, which is some 46 times faster than using **digitalWrite**.

The waveform is not very square, showing the kind of transients that you would expect at that frequency.

Another advantage of using port registers directly is that you can write to up to eight output pins simultaneously, which is very useful if you are writing to a parallel data bus.

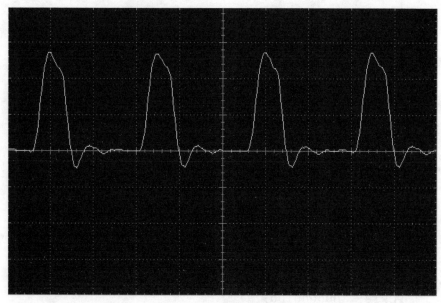

**Figure 4-5**   *Generating a 4-MHz signal with an Arduino*

## Fast Digital Input

You can also use the same method of accessing the port registers directly to speed up digital reads. Although, if you are thinking of doing this because you want to catch a very short pulse, then using interrupts is probably best (see Chapter 3).

One situation in which using the ports directly is helpful is when you want to read a number of bits simultaneously. The following sketch reads all the inputs of port B (D8 to D13) and writes the result as a binary number in the Serial Monitor (Figure 4-6).

```
// sketch_04_010_direct_read

byte state = 0;

void setup()
{
  DDRB = B00000000; // all inputs
  Serial.begin(9600);
}
```

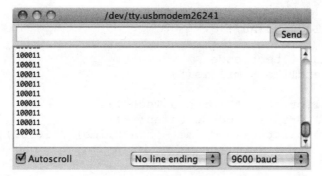

**Figure 4-6**  *Reading eight inputs at once*

```
void loop()
{
  Serial.println(PINB, 2);
  delay(1000);
}
```

The DDRB register sets all bits to 0, designating all the pins to be inputs. In the loop, you use **Serial.println** to send the number back to the Serial Monitor, where it is displayed in binary. To force it to display in binary rather than the default of decimal, use the extra **2** argument.

## Speeding Up Analog Inputs

Let's start by adapting the benchmark sketch to see just how long an **analogRead** takes before trying to speed it up:

```
// sketch 04_11_analog

void setup()
{
  Serial.begin(9600);
  while (! Serial) {};
  Serial.println("Starting Test");
  long startTime = millis();

  // test code here
  long  i = 0;
  for (i = 0; i < 1000000; i ++)
```

```
{
  analogRead(A0);
}
// end of test code
long endTime = millis();

Serial.println("Finished Test");
Serial.print("Seconds taken: ");
Serial.println((endTime - startTime) / 10001);
}

void loop()
{
}
```

This sketch takes 112 seconds to run on an Arduino Uno. That means the Uno can take nearly 9000 analog readings per second.

The **analogRead** function uses an analog-to-digital converter (ADC) in the Arduino's microcontroller. Arduino uses a type of ADC called a *successive approximation ADC*. It works by effectively closing in on the analog value by comparing it with a reference voltage that it adjusts. The ADC is controlled by a timer, and you can increase the frequency to make the conversion quicker.

The following code increases the frequency of the ADC from 128 kHz to 1 MHz, which should make things about eight times faster:

```
// sketch 04_11_analog

const byte PS_128 = (1 << ADPS2) | (1 << ADPS1) | (1 << ADPS0);
const byte PS_16 = (1 << ADPS2);

void setup()
{
  ADCSRA &= ~PS_128;   // remove prescale of 128
  ADCSRA |= PS_16;     // add prescale of 16 (1MHz)
  Serial.begin(9600);
  while (! Serial) {};
  Serial.println(PS_128, 2);
  Serial.println(PS_16, 2);
  Serial.println("Starting Test");
  long startTime = millis();

  // test code here
  long  i = 0;
  for (i = 0; i < 1000000; i ++)
```

```
{
   analogRead(A0);
}
// end of test code
long endTime = millis();

Serial.println("Finished Test");
Serial.print("Seconds taken: ");
Serial.println((endTime - startTime) / 10001);
}

void loop()
{
}
```

The code now takes only 17 seconds to run, which is roughly 6.5 times faster, increasing our samples per second to about 58,000. That is plenty fast enough to sample audio, although you won't be able to store much of it in 2kB of RAM!

If the original sketch_04_11_analog is run on an Arduino Due, the test completes in 39 seconds. You cannot use the register trick we just tried on the Due, however, as it has a different architecture.

# Summary

In this chapter, we tried to squeeze the last possible drop of juice out of our meager 16 MHz of processor power. In the next chapter, we'll switch our attention to minimizing the Arduino's power consumption, something that is quite important for battery- and solar-powered Arduino projects.

# 5

# Low Power Arduino

It is fair to say that, without taking any special measures, an Arduino really does not use a lot of power. Typically, an Arduino Uno draws about 40 mA, which when powered from USB at 5V amounts to just 200 mW. That means it can happily run on a small 9V battery (150 mAh) for perhaps four hours.

Current consumption becomes important when the Arduino is being run for long periods of time using batteries, such as in remote monitoring or control situations in which batteries or solar are the only option. For instance, I recently made an Arduino-based automatic hen-house door, using a small solar panel to charge the battery so it had enough juice to open and close the door twice a day.

## Power Consumption of Arduino Boards

Let's establish some initial figures for the power consumption of a few of the more popular Arduino boards. Table 5-1 shows the results of directly measuring the current consumption of the boards with an ammeter. Note that measuring this is a little tricky as the current varies as timers and other parts of the microcontroller and Arduino board perform periodic tasks.

One interesting thing is that if you look at the difference between an Arduino operating at 5V with and without the processor chip, the difference is just 15 mA, implying that the board itself is using the other 32 mA. The Arduino board does, of course, have the USB interface chip, an "On" LED, and 3.3V voltage regulators, all of which use some power even

| Board | Current |
|---|---|
| Uno (5V USB) | 47 mA |
| Uno (9V power supply) | 48 mA |
| Uno (5V processor removed) | 32 mA |
| Uno (9V processor removed) | 40 mA |
| Leonardo (5V USB) | 42 mA |
| Due (5V USB) | 160 mA |
| Due (9V power supply) | 70 mA |
| Mini Pro (9V power supply) | 42 mA |
| Mini Pro (5V USB) | 22 mA |
| Mini Pro (3.3V direct) | 8 mA |

**Table 5-1** *Power Consumption for Arduino Boards*

without the microcontroller. Note also how much less current the microcontroller draws at 3.3V.

The techniques described here can reduce the power required by the processor but not that required by the board itself. In the examples that follow, I use the Arduino Mini Pro board supplied directly with 3.3V through its VCC and GND connections (Figure 5-1), bypassing the voltage regulator, so that apart from the "On" LED, I am only powering the microcontroller chip.

**Figure 5-1** *An Arduino Mini Pro, powered directly from 3V*

This setup is one you would be likely to use in a battery-operated system, as a single lithium polymer (LiPo) battery cell provides 2.7V when almost empty and 4.2V when full, a range that is just fine for a naked ATmega328 microcontroller.

## Current and Batteries

This book is about software, so I will not dwell on batteries any longer than is necessary. Figure 5-2 shows a selection of batteries that you might consider for powering an Arduino.

At the top-left is a 2400mAh LiPo cylindrical LiPo battery. Below is a small, flat 850mAh LiPo battery. LiPo batteries are lightweight and can be recharged many times and hold a lot of energy for their size and weight. At the top-right is a 9V NiMh battery, with a capacity of 200 mAh. This battery is also rechargeable but uses an older technology. Because it is a 9V battery, it would be suitable for powering an Arduino only when using the Arduino's voltage regulator. You can buy battery clip adapters that allow

2400mAh LiPo battery                                                          9V NiMh battery

850mAh LiPo battery          3V Lithium battery

**Figure 5-2**  *Batteries for powering Arduino boards*

you to connect the battery to the barrel jack on an Arduino. Finally, at the bottom-right is a 3V nonrechargeable Lithium battery (CR2025) that has a capacity of about 160 mAH.

As a rule of thumb, you can calculate the number of hours that a battery will last before it is discharged by dividing the capacity in milliamp hour (mAh) by the number of milliamps (mA) being drawn:

Battery life in hours = Battery capacity in mAh / Current in mA

For example, if we were to use the CR2025 to power a Mini Pro at 3V, we could expect it to last 160mAh/8mA = 20 hours. If we powered the same hardware from the 2400 mA LiPo cell, we could expect it to last 2400/8 = 300 hours.

## Reducing the Clock Speed

Most of the Arduino family has a clock frequency of 16 MHz. The micro-controller only really uses significant amounts of current when its binary logic is switching from a HIGH to a LOW, so the frequency at which the chip operates has a big effect on the current consumed. Lowering the frequency will, of course, make the microcontroller perform more slowly, which may or may not be a problem.

You can lower the frequency at which an ATmega328 chip operates from within your sketch. A convenient way to do this is to use the Arduino Prescaler library (http://playground.arduino.cc/Code/Prescaler).

As well as allowing you to set the microcontroller's frequency of operation, the Prescaler library also provides replacement functions for **millis** and **delay** called **trueMillis** and **trueDelay**. These replacements are necessary because reducing the clock frequency will increase the length of a delay by the same proportion.

The following example sketch turns the "L" LED on for 1 second and then off for 5 seconds, during which the current is measured for each of the possible Prescaler values that set the frequency.

```
// sketch_05_01_prescale

#include <Prescaler.h>
```

```
void setup()
{
  pinMode(13, OUTPUT);
  setClockPrescaler(CLOCK_PRESCALER_256);
}

void loop()
{
  digitalWrite(13, HIGH);
  trueDelay(1000);
  digitalWrite(13, LOW);
  trueDelay(5000);
}
```

The library provides a number of constants so you can set the amount by which the clock frequency is divided. Therefore, the value **CLOCK_PRES-CALER_1** leaves the clock frequency unchanged at 16 MHz and, at the other extreme, using the constant **CLOCK_PRESCALER_256** will divide the clock frequency by 256, giving a clock frequency of just 62.5 kHz.

Table 5-2 show the current consumption at each of the possible clock frequencies, and Figure 5-3 shows these data on a chart. The chart shows that the curve starts to level off fairly steeply, so 1 MHz looks like a good compromise of clock frequency verses power consumption.

| Constant | Equivalent Clock Frequency | Current (LED off) |
|---|---|---|
| CLOCK_PRESCALER_1 | 16 MHz | 7.8 mA |
| CLOCK_PRESCALER_2 | 8 MHz | 5.4 mA |
| CLOCK_PRESCALER_4 | 4 MHz | 4.0 mA |
| CLOCK_PRESCALER_8 | 2 MHz | 3.2 mA |
| CLOCK_PRESCALER_16 | 1 MHz | 2.6 mA |
| CLOCK_PRESCALER_32 | 500 kHz | 2.3 mA |
| CLOCK_PRESCALER_64 | 250 kHz | 2.2 mA |
| CLOCK_PRESCALER_128 | 125 kHz | 2.1 mA |
| CLOCK_PRESCALER_256 | 62.5 kHz | 2.1 mA |

**Table 5-2**  *Current Consumption vs. Clock Speed*

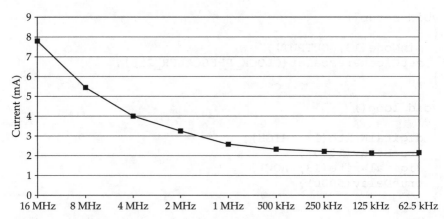

**Figure 5-3** *A chart of current consumption vs. clock speed*

As well as having to use new versions of **millis** and **delay**, there are other consequences of stopping the clock speed. In fact, any task in which timing is critical, such as PWM output and Servo control, is not going to work as expected.

Most of that 2.1 mA, used at the slowest clock speed, is likely to be consumed by the "On" LED, so if you really want to be economical, you could carefully de-solder it.

# Turning Things Off

The ATmega chips have very sophisticated power management, to the extent that you can actually turn off features that you are not using to save a small amount of current.

What is more, you can turn things on and off in your sketch. So you could, for example, just turn on the analog-to-digital converter (ADC) when you need to do an **analogRead** and then turn it off again afterward.

The power is controlled using a library **avr/power.h** that includes functions in disable/enable pairs. So the function **power_adc_disable** turns the ADC off and **power_adc_enable** turns it back on again.

The power savings to be had are not great, however. In my testing, turning everything off on a Mini Pro at 5V and 16 MHz saved a total of just

1.5 mA, reducing the current from 16.4 mA with everything on, to 14.9 with everything off. I used the following test sketch:

```
// sketch_05_02_powering_off

#include <avr/power.h>

void setup()
{
  pinMode(13, OUTPUT);
//  power_adc_disable();
  power_spi_disable();
//  power_twi_disable();
//  power_usart0_disable();
//  power_timer0_disable();
//  power_timer1_disable();
//  power_timer2_disable();
//  power_all_disable();
}

void loop()
{
}
```

The functions available are listed in Table 5-3. Each function also has a counterpart, ending in **enable** rather than **disable**.

| Function | Description |
|---|---|
| power_adc_disable | Disable analog inputs |
| power_spi_disable | Disable the SPI Interface |
| power_twi_disable | Disable TWI (I2C) |
| power_usart0_disable | Disable UART serial (serial communications over USB use this) |
| power_timer0_disable | Disable Timer 0 (**millis** and **delay** use this) |
| power_timer1_disable | Disable Timer 1 |
| power_timer2_disable | Disable Timer 2 |
| power_all_disable | Disable all the modules listed previously |

**Table 5-3**  *Power Management Functions for ATmega Arduinos*

# Sleeping

The ultimate way to save power on your Arduino is to put it to sleep when it doesn't have anything useful to do.

## Narcoleptic

Peter Knight has produced an easy-to-use library called *Narcoleptic*, which you can download from here: https://code.google.com/p/narcoleptic/.

Obviously, putting an Arduino to sleep is of no use if you can't wake it up again! There are two methods to wake up an Arduino. One is to use an external interrupt and the other is to set a timer to wake the Arduino after a period of time. The Narcoleptic library just uses the timer method.

The Narcoleptic library takes the approach of providing you with an alternative **delay** function that puts the Arduino to sleep for the time specified in the **delay**. Because nothing happens when the Arduino is doing a **delay** anyway, this method works brilliantly.

For example, let's look at our old favorite sketch, Blink. The following sketch turns an LED on for 1 second and then turns it off for 10 seconds and repeats indefinitely:

```
// sketch_05_03_blink_standard

void setup()
{
  pinMode(13, OUTPUT);
}

void loop()
{
  digitalWrite(13, HIGH);
  delay(1000);
  digitalWrite(13, LOW);
  delay(10000);
}
```

The Narcoleptic version of this sketch is shown here:

```
// sketch_05_04_narcoleptic_blink
#include <Narcoleptic.h>
```

```
void setup()
{
  pinMode(13, OUTPUT);
}

void loop()
{
  digitalWrite(13, HIGH);
  Narcoleptic.delay(1000);
  digitalWrite(13, LOW);
  Narcoleptic.delay(10000);
}
```

The only difference is that you import the Narcoleptic library and use its **delay** rather than the regular **delay**.

Running both sketches on a Mini Pro at 5V and 16 MHz, the first sketch uses around 17.2 mA when the LED is in the off part of the cycle. On the other hand, the Narcoleptic version of the sketch reduces this to a tiny 3.2 mA. The "On" LED uses most of that (about 3 mA), so if you remove it, then your average power consumption could be reduced to well under 1 mA.

The microcontroller can go to sleep pretty quickly, so if your project relies on a button being pressed to trigger some action, you do not necessarily need to use an external interrupt to wake it from sleep. But you could (probably more easily) write your code so the Arduino wakes 10 times a second, checks to see if an input is HIGH, and then, if it is, does something rather than go back to sleep. The following sketch illustrates this process:

```
// sketch_05_05_narcoleptic_input
#include <Narcoleptic.h>

const int ledPin = 13;
const int inputPin = 2;

void setup()
{
  pinMode(ledPin, OUTPUT);
  pinMode(inputPin, INPUT_PULLUP);
}
```

```
void loop()
{
  if (digitalRead(inputPin) == LOW)
  {
    doSomething();
  }
  Narcoleptic.delay(100);
}

void doSomething()
{
  for (int i = 0; i < 20; i++)
  {
    digitalWrite(ledPin, HIGH);
    Narcoleptic.delay(200);
    digitalWrite(ledPin, LOW);
    Narcoleptic.delay(200);
  }
}
```

When running this sketch, a Mini Pro at 5V and 16 MHz uses a miserly 3.25 mA while the Arduino waits for something to happen. When pin2 is connected to ground, the LED is flashed 20 times, but because you are using the Narcoleptic **delay** in the LED flashing too, the current only rises to an average of 4 or 5 mA.

If you change the **delay** inside the loop, to try and make the Arduino wake, say, 100 times per second, the power will rise again because it does take a little while for the Arduino to go to sleep. A delay of 50 (20 times a second), however, would work just fine.

## Waking on External Interrupts

The approach just described works for most situations; however, if you need to respond more quickly to an external event, then you need to arrange for the microcontroller to wake up when an external interrupt occurs.

To rework the previous example to use pin D2 as an external interrupt pin is a lot more work, but it achieves slightly better results, as it does not require polling the interrupt pin. The code for this is quite complex, so first I'll show you the code and then describe how it all works. If you skipped Chapter 3 on interrupts, then you should probably read it before tackling this example.

```
// sketch_05_06_sleep_external_wake
#include <avr/sleep.h>

const int ledPin = 13;
const int inputPin = 2;

volatile boolean flag;

void setup()
{
  pinMode(ledPin, OUTPUT);
  pinMode(inputPin, INPUT_PULLUP);
  goToSleep();
}

void loop()
{
  if (flag)
  {
    doSomething();
    flag = false;
    goToSleep();
  }
}

void setFlag()
{
  flag = true;
}

void goToSleep()
{
  set_sleep_mode(SLEEP_MODE_PWR_DOWN);
  sleep_enable();
  attachInterrupt(0, setFlag, LOW); // pin D2
  sleep_mode(); // sleep now
  // Now asleep until LOW interrupt, then..
  sleep_disable();
  detachInterrupt(0);
}
```

```
void doSomething()
{
  for (int i = 0; i < 20; i++)
  {
    digitalWrite(ledPin, HIGH);
    delay(200);
    digitalWrite(ledPin, LOW);
    delay(200);
  }
}
```

The first thing to note is that the example uses some functions that are defined in the library **avr/sleep.h**. Just like **avr/power.h** that I used earlier, this library is not part of the Arduino core, but rather a library for the AVR family of microcontrollers. This means it will not work on the Arduino Due, but then again, if you are making a low-power Arduino project, the Due should be just about your last choice of board.

After defining the pins I am going to use, I then define a **volatile** variable to allow the ISR to communicate with the rest of the sketch.

The **setup** function sets up the pins and then calls the function **goToSleep**. This function sets the type of sleep mode, which, in this case, is **SLEEP_MODE_PWR_DOWN**. This mode saves the most power, so it makes sense to use it.

It is then necessary to call **sleep_enable**. Calling this does not actually put the microcontroller to sleep. Before I do that, I need to attach an interrupt to interrupt 0 (pin D2) so the Arduino can be woken when the time comes.

*NOTE*    *Notice that the interrupt type is set to* **LOW***. This is the only interrupt type that you can use with this sleep example. RISING, FALLING, and CHANGE will not work.*

Having attached the interrupt, calling **sleep_mode()** actually puts the process to sleep. When the microcontroller eventually wakes, the ISR is run and then the sketch continues from the next line in **goToSleep**. This first calls **disable_sleep** and then detaches the interrupt, so the ISR cannot be invoked again until the sketch has put itself back to sleep.

When an interrupt occurs on D2, the ISR (**setFlag**) simply sets a flag that the **loop** function checks. Remember that using **delay**s and so on, in an ISR are a no-no. The **loop** function must, therefore, monitor the flag until it becomes set and then call the same **doSomething** function that was used in the Narcoleptic example. Having performed the action, the flag is reset and the Arduino put back to sleep.

The power consumption level was pretty much the same as in the Narcoleptic example, except that while flashing the LEDs, the current consumption was higher as the normal **delay** function was used.

# Use Digital Outputs to Control Power

Although this chapter is really about using software to minimize power consumption, it would not be out of place to mention a useful hardware tip to keep the power consumption low.

Figure 5-4 shows a light sensor using a photoresistor (resistance changes with light) and a fixed resistor connected to an Arduino analog input that is measuring the light intensity.

The problem with this approach is that there is a constant current flowing from 5V through the photoresistor and then through the fixed resistor. If the photoresistor has a "bright" resistance of 500Ω, then, using Ohm's Law, the current flowing is $I = V/R = 5V / (1000Ω + 500Ω) = 3.3$ mA.

Instead of using the fixed 5V supply of the Arduino, you could use a digital output (see Figure 5-5) to turn the pin HIGH, take a reading, and then turn it LOW again. In this way, the 3.3 mA only flows for a tiny amount of time every time a reading is taken, reducing the average current consumption enormously.

The following sketch illustrates this approach:

```
// sketch_05_07_light_sensing

const int inputPin = A0;
const int powerPin = 12;

void setup()
{
  pinMode(powerPin, OUTPUT);
  Serial.begin(9600);
}
```

```
void loop()
{
  Serial.println(takeReading());
  delay(500);
}

int takeReading()
{
  digitalWrite(powerPin, HIGH);
  delay(10); // photoresistors are slow to respond
  int reading = analogRead(inputPin);
  digitalWrite(powerPin, LOW);
  return reading;
}
```

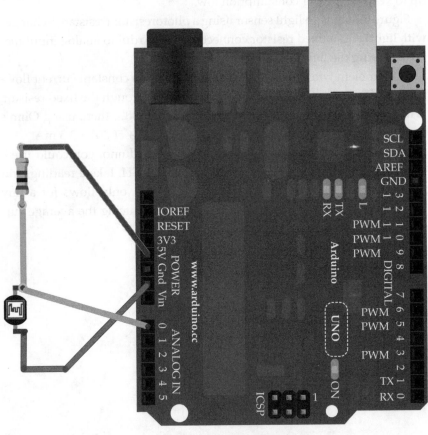

**Figure 5-4** *Measuring light with an LDR (Photoresistor)*

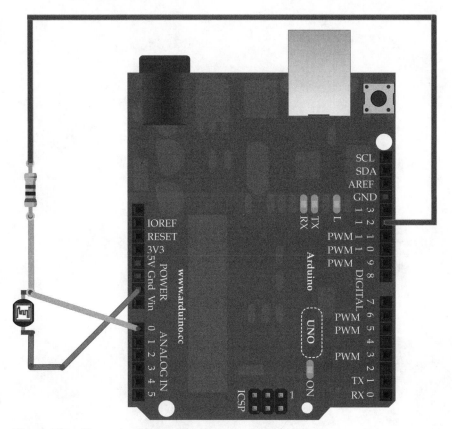

**Figure 5-5**  *Measuring light economically*

You can use this technique for a lot more than just light sensing. You could, for example, use the digital output to use a MOSFET transistor to turn high-power parts of your project on and off as required.

## Summary

The best ways to minimize current consumption are to:

- Put the microcontroller to sleep when it's not doing anything
- Run the Arduino at a lower voltage
- Run the Arduino at a lower clock frequency

# 6
## Memory

**Whereas most** computers have memory capacities measured in gigabytes, the Arduino Uno has just 2kB. That is more than a million times less memory than a conventional computer. Having only a little memory to work with focuses the mind wonderfully when writing code, however. There is no room for the "bloatware" that plagues most computers.

Although writing memory-efficient code is important, you shouldn't do so at the expense of writing code that is easy to read and maintain. Even with an Arduino's limited resources, most sketches will not get close to using all the RAM. You really only need to worry about memory capacity when you have a very complex sketch or a sketch that uses a lot of data.

## Arduino Memory

Comparing an Arduino's memory with that of conventional computers is a little unfair, as they actually use their RAM memory in different ways. Figure 6-1 shows how a PC uses its memory when running a program.

When a PC runs a program, it first copies the entire program from the hard disk into RAM and then executes that copy of the program. Variables in the program then use more of the RAM. By contrast, Figure 6-2 shows how an Arduino uses memory when a program is run. The program itself actually runs directly from flash memory. It is not copied into RAM.

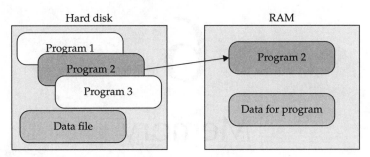

**Figure 6-1** *How a PC uses memory*

The RAM in an Arduino is only used to hold the contents of variables and other data relating to the running of the program. RAM is not persistent; that is, when the power is disconnected, the RAM is cleared. If the program needs to store persistent data, then it must write that data to EEPROM. The data can then be read back when the sketch restarts.

When pushing the limits of an Arduino, you have to worry about both RAM usage and, to a lesser extent, the size of the program in flash memory. Because an Arduino Uno has 32kB of flash, this limit is not often reached.

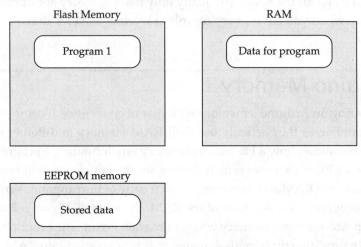

**Figure 6-2** *How an Arduino uses memory*

# Minimizing RAM Usage

As you have seen, the way to reduce RAM usage is to reduce the amount of RAM used by variables.

## Use the Right Data Structures

By far, the most common data type in Arduino C is the **int**. Each **int** uses 2 bytes, but most of the time, you don't represent a number between –32,768 and +32,767, and the much smaller range of 0 to 255 offered by a "byte" does just fine. Most built-in methods that work with an **int**, will work just the same with a byte.

A common example of how this works is variables used for pin numbers. It is common to use **int**s for this, as shown in the following example:

```
// sketch_06_01_int
int ledPins[] = {2, 3, 4, 5, 6, 7, 8, 9, 10, 11, 12, 13};

void setup()
{
  for (int i = 0; i < 12; i++)
  {
    pinMode(ledPins[i], OUTPUT);
    digitalWrite(ledPins[i], HIGH);
  }
}

void loop()
{
}
```

You could easily change the **int** array to be an array of bytes instead. If you do this, the program functions just the same time, but the array will occupy half the memory.

A really good way to reduce memory usage is to make sure that any constant variables are declared as such. To do this, just put the word **const** in front of the variable declaration. Knowing that the value will never change allows the compiler to substitute in the value in place of the variable, which saves space. For example, the array declaration in the previous example becomes

```
const byte ledPins[] = {2, 3, 4, 5, 6, 7, 8, 9, 10, 11, 12, 13};
```

## Be Careful with Recursion

*Recursion* is a technique where a function calls itself. Recursion can be a powerful way of expressing and solving a problem. In functional programming languages such as LISP and Scheme, recursion is used a great deal.

When a function is called, an area of memory called the *stack* is used. Imagine a spring-loaded sweet dispenser, like a Pez™ dispenser, except you are only going to push sweets in from the top or pop sweets off from the top (Figure 6-3). The term *push* is used to indicate something being added to the stack and *pop* indicates taking something off the stack.

Every time you call a function, a stack frame is created. A *stack frame* is a small memory record that includes storage space for parameters and local variables used by the function, as well as a return address that specifies the point in the program from which execution should continue when the function has finished running and returned.

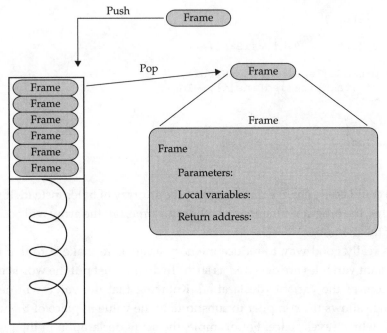

**Figure 6-3**    *The stack*

Initially, the stack is empty, but when you call a function (let's call it "function A"), memory is allocated for it and a stack frame is pushed onto the stack. If function A calls another function (function B), then a second record is added to the top of the stack, so the stack now has two records. When function B finishes, its stack frame is popped off the stack, then when function A completes, its stack frame is also popped off the stack. Because local variables for a function are stored on the stack frame, they are not remembered between successive function calls.

The stack uses some of our valuable memory, and most of the time, the stack never has more than three or four records on it. The exception is if you allow functions to call themselves, or to be in a loop of functions calling each other. Then there is the real possibility that the program will run out of stack memory.

For example, the mathematical "factorial" function is calculated by multiplying all the integers that come before a number up to that number. The factorial of 6 is $6 \times 5 \times 4 \times 3 \times 2 \times 1 = 720$.

A recursive definition of the factorial of $n$ is

If $n = 0$, the factorial of $n$ is 1.

Otherwise the factorial of $n$ is $n$ times the factorial of $(n - 1)$.

You can write this in Arduino C as:

```
long factorial(long n)
{
  if (n == 0)
  {
    return 1;
  }
  else
  {
    return n * factorial(n - 1);
  }
}
```

You can find this code and a full sketch that prints the result in **sketch_06_02_factorial**. Generally, people with a mathematical mind think this is pretty neat. You'll notice, however, that the depth of the stack equals the number of the factorial you are calculating. It is also pretty easy to see how to write a nonrecursive version of the **factorial** function:

```
long factorial(long n)
{
  long result = 1;
  while (n > 0)
  {
    result = result * n;
    n--;
  }
  return result;
}
```

In terms of ease of reading, this code is probably easier to understand; it also uses less memory and is faster. In general, it makes sense to avoid recursion, or restrict it to highly efficient recursive algorithms like Quicksort (http://en.wikipedia.org/wiki/Quicksort), which can put an array of numbers into order very efficiently.

## Store String Constants in Flash Memory

By default, if you declare string constants as shown in the following example, those character arrays will be stored in RAM and in flash memory— once for the program code and once when their values are copied into RAM when the sketch is run:

```
Serial.println("Program Started");
```

If, however, you use the following code, the string constant will be stored in flash memory only:

```
Serial.println(F("Program Started"));
```

In the "Using Flash" section in this chapter, you'll see how you can use flash in other ways.

## Common Misconceptions

A common misconception is that using short variable names uses less memory. This is not the case. The compiler takes care of such things, so the final variable names do not find their way into the binary sketch. Another misconception is that comments in a program have an effect on the size of the program when it is installed or on the RAM that it uses. This is not true.

You may also assume that dividing your code into lots of small functions will increase the size of the compiled code. This is not usually the case as the compiler is smart enough to actually replace function calls with inline copies of the body of the function as part of its code optimization process. This benefit allows you to write more readable code.

## Measure Free Memory

You can find out how much RAM a running sketch is using at any point in time with the **MemoryFree** library, which you can download from here: http://playground.arduino.cc/Code/AvailableMemory.

This library is easy to use; it provides a function called **freeMemory**, which returns the number of bytes available. The following sketch illustrates its use:

```
#include <MemoryFree.h>

void setup()
{
    Serial.begin(115200);
}

void loop()
{
    Serial.print("freeMemory()=");
    Serial.println(freeMemory());
    delay(1000);
}
```

This library can be handy if you start to experience unexplained problems with a sketch that you think might be caused by a memory shortage. The library does, of course, increase your memory usage a little.

# Minimizing Flash Usage

When you successfully compile a sketch, you'll see a status line at the end of the process that says something like this:

```
Binary sketch size: 1,344 bytes (of a 32,256 byte maximum)
```

This line tells you exactly how much of the Arduino's flash memory the sketch will use, so you know if you're getting close to the 32kB limit. If you are not near the limit, then you don't really need to try to optimize the flash memory. If you are getting close, then there are a few things that you can do.

## Use Constants

When variables are defined, especially pin names, it is quite common to see them defined like this:

```
int ledPin = 13;
```

Unless you plan to change which pin is to be used as the LED pin while the sketch is actually running, then you can use a constant. Just add the word **const** to the front of the declaration:

```
const int ledPin = 13;
```

This change saves you 2 bytes, plus 2 bytes for every place that the constant is used. For a much used variable, your savings can amount to a few tens of bytes.

## Remove Unwanted Trace

When debugging Arduino sketches, sprinkling the code with **Serial.println** commands helps you see the value of variables and work out any bugs in the program. These commands actually use a fair bit of flash memory. Any use of **Serial.println** pulls about 500 bytes of library code into the sketch. So, once you are convinced that the sketch is working, remove or comment out these lines.

## Bypass the Bootloader

Back in Chapter 2, you discovered how to program the microcontroller directly on the Arduino using the ISP connector and programming hardware. This approach can save you a valuable couple of kBs, as it means the bootloader does not need to be installed.

# Static vs. Dynamic Memory Allocation

If, like the author, you come from a background of writing large-scale systems in languages such as Java or C#, you're used to creating objects at runtime and allowing a garbage collector to tidy up behind you. This approach to programming is simply inappropriate on a microprocessor with just 2kB of memory. For a start, there is simply no garbage collector, and what is more, allocating and deallocating memory at runtime is rarely necessary in the type of programs written for an Arduino.

The following example defines an array statically, as you would normally in a sketch:

```
// sketch_06_04_static

int array[100];

void setup()
{
   array[0] = 1;
   array[50] = 2;
   Serial.begin(9600);
   Serial.println(array[50]);
}

void loop()
{
}
```

The memory that the array uses is known while the sketch is being compiled; therefore, the compiler can reserve the necessary amount of memory for the array. This second example also creates an array of the same size, but it allocates the memory for it at runtime, from a pool of available memory. Note that versions of the Arduino software prior to 1.0.4 do not support **malloc**.

```
// sketch_06_05_dynamic

int *array;

void setup()
{
  array = (int *)malloc(sizeof(int) * 100);
  array[0] = 1;
  array[50] = 2;
  Serial.begin(9600);
  Serial.println(array[50]);
}

void loop()
{
}
```

You start by defining a variable *int \*array*. The * indicates that this is a pointer to an integer value (or, in this case, array of ints) rather than a simple value. The memory to be used by the array itself is not claimed for use by the array until the following line is executed in **setup**:

```
array = (int *)malloc(sizeof(int) * 100);
```

The **malloc** (*memory allocate*) command allocates memory from an area of RAM called the *heap*. Its argument is the number of bytes to be allocated. Because the array contains 100 **int**s, you need to do a little calculation to work out how many bytes to reserve. Actually, you could just write **200** as the parameter to **malloc** because you know that each **int** occupies 2 bytes of memory, but by using the **sizeof** function, you can make sure to get the right number.

After the memory has been allocated, you can use the array just as if you had allocated it statically. The only advantage to allocating it dynamically is that you can delay the decision about how large to make it until the sketch is actually running (runtime).

The danger with dynamic memory allocation is that you can easily get in a situation where memory is allocated but not released, so then the sketch unexpectedly runs out of memory. Running out of memory can cause the Arduino to hang. If all the memory is allocated statically, however, this cannot happen.

Note that I have developed hundreds of Arduino projects and have yet to find a compelling reason to use dynamic memory allocation on an Arduino.

# Strings

*Strings* (text) are used less commonly in Arduino programming than in more conventional software development. In most software development, strings are the most used data type because most programming is about user interfaces or databases, which naturally involve text of some sort.

Many Arduino programs have no need to represent strings of text at all, or, if they do, it's in **Serial.println** commands used to debug a buggy sketch.

There are essentially two methods for using strings in Arduino: the old way (C **char** arrays) and the new way, the String Object library.

## C char Arrays

When you define a string constant by typing something like

```
char message[] = "Hello World";
```

you are statically defining a **char** array that is 12 characters long. It is 12 characters rather than the 11 letters of "Hello World" because there is a final terminating character of 0 to mark the end of the string. This is the convention for C character strings, and it allows you to use larger arrays of characters than the string that you are interested in at the start (Figure 6-4). Each character letter, number, or other symbol has a code called its ASCII value.

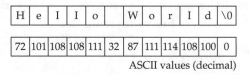

ASCII values (decimal)

**Figure 6-4**  *A null-terminated C char array*

Note that another commonly used convention for string constants is to write:

```
char *message = "Hello World";
```

This syntax works similarly but declares **message** to be a pointer to a character (the first character of the array).

## Formatting Strings with Multiple Prints

Much of the time, this is the only way you need to use a string, for instance, to display a message on an LCD screen or as a parameter to **Serial.println**. You may think that being able to join strings and convert numbers to strings is essential. For example, let's look at a specific problem—how to display a message on an LCD screen such as "Temp: 32 C." You might believe you need to join the number 32 to a string **"Temp: "** and then add the string **" C"** onto the end. Indeed, if you are a Java programmer, you will probably expect to write the following in C:

```
String text = "Temp: " + tempC + " C";
```

Sorry, that's not the way it works in C. In this case, you can print this message simply by using multiple **print** statements, as shown in this example:

```
lcd.print("Temp: "); lcd.print(tempC); lcd.print(" C");
```

This method removes the need for any behind-the-scenes copying of data that would go on during string concatenation in other newer languages.

The same multiple outputs approach works with the Serial Monitor and **Serial.print** statements. In this case, you generally make the last print on the line a **println** to add a newline to the output.

## Formatting Strings with sprintf

The standard C string library (not to be confused with the Arduino String Object library discussed in the next section) includes a very useful function called **sprintf** to format character arrays. This fits variables into a pattern string, as shown in the following example:

```
char line1[17];
int tempC = 30;
sprintf(line1, "Temp: %d C", tempC);
```

The character array line1 is a string buffer that is used to contain the formatted text. The size is specified as 17 to allow an extra null character on the end. I chose the name line1 to illustrate how this could be the contents of the top line of a 16-character by two-line LCD display.

The sprintf command's first parameter is the character array into which the result is to be written. The next argument is the formatting string that contains a mixture of literal text like Temp: and formatting commands like %d. In this case, %d means signed decimal. The remainder of the parameters will be substituted in order into the formatting string in place of the formatting commands.

If your LCD display were to show the time on the second line, then you could format the time from separate hours, minutes, and seconds using the following line:

```
char line2[17];
int h = 12;
int m = 30;
int s = 5;
sprintf(line2, "Time: %2d:%02d:%02d", h, m, s);
```

If you were to print line2 to the Serial Monitor or an LCD screen, it would look like this:

```
Time: 12:30:05
```

Not only have the numbers been substituted in the correct place, but also a leading zero is in front of the 5 digit. In the sketch, between each : you have the formatting commands for the three parts of the time. For the hour, it is %2d, which means display the value with a length of two digits as a decimal. The formatting command for minutes and seconds is slightly different (%02d). This command still means format as two characters, but include a leading zero.

Be wary, though, this approach works for ints, but the Arduino developers have not implemented the standard C library formatting for other types such as floats.

## Finding the Length of a String

Because the string within a character array is often smaller than the actual character array containing it, a useful function, called strlen, is available.

**strlen** counts the number of characters in the array before the null that marks the end of the string.

The function returns the size of the string (excluding the null) and takes the character array as its only argument, for instance,

```
strlen("abc")
```

returns the number 3.

# The Arduino String Object Library

Versions of the Arduino IDE since version 019, several years ago, have included a String library that is far more familiar and friendly to developers used to Java, Ruby, Python, and so on, where the norm is to construct strings by concatenation, often using "+". This library also offers a whole host of useful string searching and manipulation features.

This library, of course, comes at the cost of adding several kBs to your sketch size should you use it. It also uses dynamic memory allocation, with all its associated problems of running out of memory. So think carefully before you decide to use it. Many Arduino users stick to C character arrays instead.

This library is beautifully easy to use, and if you have used strings in Java, you will be very at home with the Arduino String Object library.

## Creating Strings

You can create the string using a **char** array, **int**, or **float**, as shown in the following example:

```
String message = "Temp: ";
String temp = String(123);
```

## Concatenating Strings

Strings can then be concatenated with each other and other data types using +. Try placing the following code in the **setup** function of an otherwise empty sketch:

```
Serial.begin(9600);
String message = "Temp: ";
String temp = String(123);
Serial.println(message + temp + " C");
```

| Function | Example | Description |
|---|---|---|
| [] | `char ch = String("abc")[0]` | ch gets the value **a**. |
| trim | `String s = "    abc    ";`<br>`s.trim();` | Removes the space characters either side of **abc**.<br>**s** is left with the value **abc**. |
| toInt | `String s = "123";`<br>`int x = s.toInt();` | Converts the number in the string to an **int** or a **long** |
| substring | `s = "abcdefg";`<br>`String s2 = s.substring(1, 3));` | Returns a section of the original string. **s2** has the value **bc**. Parameters are from index and before index. |
| replace | `String s = "abcdefg";`<br>`s.replace("de", "DE");` | Replaces all occurrences of **"de"** with **"DE"** in the string. **s** is left with the value of **"abcDEfg"**. |

**Table 6-1**   *Some Useful String Functions*

Notice how the final value being concatenated to the **String** is actually a character array. As long as the first item in the sequence of values in between the + signs is a string, the items will automatically be converted into strings before being concatenated.

## Other String Functions

Table 6-1 summarizes some of the more useful things that you can do with String functions. For chapter and verse on the functions available, see this reference: http://arduino.cc/en/Reference/StringObject.

# Using EEPROM

The contents of any variable used in an Arduino sketch will be cleared and lost whenever the Arduino loses power or is reset. If you need to store values persistently, you need to write them a byte at a time into EEPROM memory. The Arduino Uno has 1kB of EEPROM memory.

**NOTE**   *This is not an option for the Arduino Due, which does not have any EEPROM. Instead you must write data to a microSD card.*

Reading and writing to EEPROM memory requires a library that is pre-installed in the Arduino IDE. The following example shows how to write a single byte of EEPROM, in this case, from the **setup** function:

```
#include <EEPROM.h>
void setup()
{
  byte valueToSave = 123;
  EEPROM.write(0, valueToSave);
}
```

The first argument of the **write** function is the address in the EEPROM to which the byte should be written, and the second argument is the value to be written to that address.

The **read** command is used to read the data back from EEPROM. To read back a single byte, you just use the line

```
EEPROM.read(0);
```

where **0** is the EEPROM address.

## EEPROM Example

The following example shows a typical scenario where a value is written during the normal running of a program and then read back during startup. The application is a door lock project using the Serial Monitor to enter codes and change the secret code. The EEPROM is used so the secret code can be changed. If the code had to be reset every time the Arduino started, then there would be no point in allowing the user to change the code.

During the discussion that follows, certain areas of the sketch will be highlighted. If you wish to see the entire sketch in your Arduino IDE, it is called **sketch_06_06_EEPROM_example** and can be found with the rest of the code for this book at www.simonmonk.org. You may find it useful to run the sketch to get a feel for how it works. It does not require that you connect any extra hardware to the Arduino.

The **setup** function contains a call to the function **initializeCode**.

```
void initializeCode()
{
  byte codeSetMarker = EEPROM.read(0);
  if (codeSetMarker == codeSetMarkerValue)
```

```
{
  code = readSecretCodeFromEEPROM();
}
else
{
  code = defaultCode;
}
}
```

This function's job is to set the variable *code* (the secret code) to its value. This value is generally a value read from EEPROM, but there are a few difficulties with this setup.

EEPROM contents are not cleared by uploading a new sketch; once written, EEPROM values can only be changed by writing a new value on top of the old value. So if this is the first time that the sketch has been run, then there is no way to know what value might be left in EEPROM by a previous sketch. You could be left with a lock, that is, a code whose value you do not know.

One way around this is to create a separate sketch specifically to set the default code. This sketch would need to be installed on the Arduino before the main sketch.

A second, less reliable, but more convenient approach is to use a marker value that you write to the EEPROM to indicate that the EEPROM has had a code written to it. The downside of this approach is there is a slim chance that the EEPROM location used to store this flag already contains it. If so, this solution would be unacceptable if you were defining a commercial product, but here you can elect to take that risk.

The **initializeCode** function reads the first byte of EEPROM and if it equals **codeMarkerValue**, which is set elsewhere to 123, it is assumed that the EEPROM contains the code and the function **readSecretCodeFromEE-PROM** is called:

```
int readSecretCodeFromEEPROM()
{
  byte high = EEPROM.read(1);
  byte low = EEPROM.read(2);
  return (high << 8) + low;
}
```

This function reads the 2-byte int code in bytes 1 and 2 of the EEPROM (Figure 6-5).

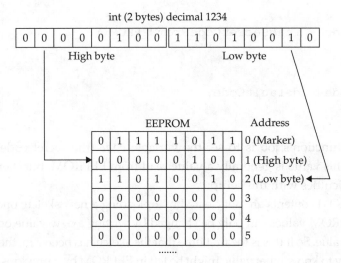

**Figure 6-5** *Storing an int in EEPROM*

To convert the two separate bytes into a single **int**, you have to shift the high bytes to the right 8 binary digits (high << 8) and then add the low bytes.

The stored code is only read when the Arduino resets. You should, however, write the secret code to EEPROM every time it is changed, so if the Arduino is powered down or reset, it still has the code available in EEPROM to be read back.

The function **saveSecretCodeToEEPROM** is responsible for this:

```
void saveSecretCodeToEEPROM()
{
  EEPROM.write(0, codeSetMarkerValue);
  EEPROM.write(1, highByte(code));
  EEPROM.write(2, lowByte(code));
}
```

This sets the code marker in EEPROM position 0 to indicate that there is a valid code in EEPROM and then writes the two bytes of the code to EEPROM. The Arduino utility functions **highByte** and **lowByte** are used to separate the parts of the **int** code.

# Using the avr/eeprom.h Library

The Arduino EEPROM library only allows you to read and write one byte at a time. In the example shown in the previous section, you got around this restriction by splitting the **int** into two bytes in order to save and retrieve it in EEPROM. An alternative is to use the underlying EEPROM library provided by AVR. This gives you more options, including reading and writing a Word (16 bits) and blocks of memory of arbitrary size.

The following sketch uses this library to save and read an **int** directly, incrementing it every time the Arduino restarts:

```
// sketch_06_07_avr_eeprom_int

#include <avr/eeprom.h>

void setup()
{
  int i = eeprom_read_word((uint16_t*)10);
  i++;
  eeprom_write_word((uint16_t*)10, i);
  Serial.begin(9600);
  Serial.println(i);
}

void loop()
{
}
```

The argument to **eeprom_read_word** (10) and the first argument to **eeprom_write_word** are the starting position of the word. Note that this occupies two bytes, so if you want to save another **int**, you specify an address of 12, not 11. The text **(uint16_t*)** before 10 is needed to make the index position the type expected by the library function.

The other useful pair of functions in this library are **eeprom_read_block** and **eeprom_write_block**. These functions allow data structures of any length (space permitting) to be stored and retrieved.

For example, let's make a sketch to write a character array string, starting at position 100 in EEPROM:

```
// sketch_06_07_avr_eeprom_string

#include <avr/eeprom.h>
```

```
void setup()
{
  char message[] = "I am written in EEPROM";
  eeprom_write_block(message, (void *)100,
        strlen(message) + 1);
}

void loop()
{
}
```

The first argument to **eeprom_write_block** is the pointer to the char array to be written, the second is the starting location in EEPROM (**100**). The final argument is the number of bytes to write. This is calculated here as the length of the string plus one to include the null character at the end of the string.

The following sketch reads the string back in again and displays it on the Serial Monitor along with the string length:

```
// sketch_06_07_avr_eeprom_string_read

#include <avr/eeprom.h>

void setup()
{
  char message[50]; // big enough
  eeprom_read_block(&message, (void *)100, 50);
  Serial.begin(9600);
  Serial.println(message);
  Serial.println(strlen(message));
}

void loop()
{
}
```

To read the string, a character array of size **50** is created. The function **eeprom_read_block** is then used to read the next 50 characters into **message**. The & sign before **message** provides the function with the message's address in RAM.

Because the message has a null on the end, when it is printed by the Serial Monitor, only the text expected (not the full 50 characters) is displayed.

# EEPROM Limitations

EEPROM is slow to read and write (about 3ms). It is also only guaranteed to be reliable for 100,000 write cycles before it starts suffering from amnesia. For this reason, you need to be careful not to write to it every time around a loop, for example.

# Using Flash

An Arduino has a lot more flash memory than it does any other type of memory. For an Arduino Uno, that is 32kB compared with 2kB of RAM. This makes it a tempting place to store data, especially as flash memory does not forget when it loses power.

There are, however, a few snags with storing data in flash memory:

- The flash memory in an Arduino can only be written to about 10,000 times before it becomes useless.

- The flash contains your program, so, if you miscalculate and write over the program, very strange things could happen.

- The flash also contains the bootloader and overwriting that will "brick" your Arduino unless you have an ISP programmer to rescue it (see Chapter 2).

- Flash can only be written a block (64 bytes) at a time.

Having said all that, it is quite easy and safe to use flash to hold constant data that are not going to change during the running of a sketch.

A third-party library is being developed that allows the Arduino Due's flash memory to be read and written to, to make up for its lack of EEPROM. You can find out more about this project here: http://pansenti.wordpress .com/2013/04/19/simple-flash-library-for-arduino-due/.

The easiest way to create flash-stored string constants is to use the **F** function that I described in an earlier section. The syntax is repeated here as a reminder:

```
Serial.println(F("Program Started"));
```

This form only works when you are using the string constant directly in a message like this. You cannot, for example, assign the result to a **char** pointer.

A more flexible, and therefore more complex, way of doing this is to use the Program Memory (PROGMEM) directive, which can be used to store any data structure. The data, however, must be constant data that will not change during the running of the sketch.

The following example illustrates how you can create an array of **ints** that will be stored in flash memory:

```
// sketch_06_10_PROGMEM_array

#include <avr/pgmspace.h>

PROGMEM int value[] = {10, 20, 25, 25, 20, 10};

void setup()
{
  Serial.begin(9600);
  for (int i = 0; i < 6; i++)
  {
    int x = pgm_read_word(&value[i]);
    Serial.println(x);
  }
}

void loop()
{
}
```

By putting the **PROGMEM** directive in front of the array declaration, you ensure that it is only stored in flash memory. To read value out of it, however, you now have to use the function **pgm_read_word** from the **avr/ pgmspace** library:

```
  int x = pgm_read_word(&value[i]);
```

The parameter to this function uses the **&** symbol in front of the array name to indicate that it is the address of this array element in flash memory that is required rather than the value itself.

The **pgm_read_word** function reads a word (2 bytes) from flash; you can also use the **pgm_read_byte** and **pgm_read_dword** to read 1 byte and 4 bytes, respectively.

# Using SD Card Storage

Although Arduino boards do not have SD card slots, several different types of shield, including the Ethernet shield and the MP3 shield shown in Figure 6-6, do have an SD or microSD card slot.

SD cards use the SPI bus interface (the topic of Chapter 9). Fortunately, to use SD cards with Arduino, you do not need to do any low-level SPI programming as there is a library included with the Arduino IDE called simply "SD."

This library includes a set of example sketches for using the SD card in various ways, including finding out information about the SD card as displayed in the Serial Monitor, as shown in Figure 6-7.

**Figure 6-6**   *MP3 shield with microSD card slot*

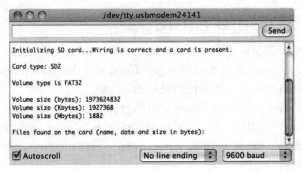

**Figure 6-7**   *Results of the Cardinfo example sketch*

Writing to the SD card is made easy, as the code snippet here shows:

```
File dataFile = SD.open("datalog.txt", FILE_WRITE);

// if the file is available, write to it:
if (dataFile) {
  dataFile.println(dataString);
  dataFile.close();
  // print to the serial port too:
  Serial.println(dataString);
}
```

# Summary

In this chapter, you have learned about all aspects of memory and data storage within Arduino. In the next chapters, you will explore Arduino programming for various types of serial interface, starting with the I2C bus.

# 7

# Using I2C

The I2C (pronounced "I squared C") interface bus is a standard for connecting microcontrollers and peripherals together. I2C is sometimes referred to as Two Wire Interface (TWI). All the Arduino boards have at least one I2C interface to which you can attach a wide range of peripherals. Some examples are shown in Figure 7-1.

The three devices on the top row of Figure 7-1 are all display modules from Adafruit. On the bottom row, starting on the left, is a TEA5767 FM receiver module. You can find these modules on eBay and elsewhere for a few dollars. The TEA5767 provides you with a full FM receiver module that you can tune to a certain frequency by sending it I2C commands. In the center is a real-time clock (RTC) module, including an I2C chip and crystal oscillator that maintains a fairly accurate time and date. Once you have set the date and time over I2C, you can read the time and date back over I2C whenever you need it. This module also includes a long-life lithium button cell that allows it to keep time, even when the module has no external power. Finally, on the right, is a 16-channel servo/PWM driver that can give you 16 extra analog outputs from your Arduino.

The I2C standard is defined as a "bus" standard because its use is not limited to connecting one component directly to another. Say you have a display connected to a microcontroller; using the same two bus pins, you can connect a whole set of "slave" devices to a "master" device. The Arduino acts as the "master," and each of the "slaves" has a unique address that identifies the device on the bus.

Adafruit 7-segment
LED display

Adafruit LED
matrix

Small Adafruit LED
matrix

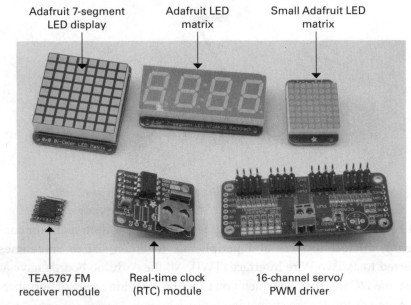

TEA5767 FM
receiver module

Real-time clock
(RTC) module

16-channel servo/
PWM driver

**Figure 7-1** *A collection of I2C devices*

Figure 7-2 shows a possible arrangement of two I2C components attached to an Arduino, a real-time clock (RTC), and a display module.

You can also use I2C to connect two Arduinos together so they can exchange data. In this case, one of the Arduinos will be configured to act as a "master" and one as a "slave."

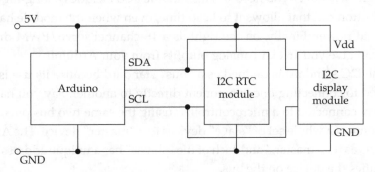

**Figure 7-2** *An Arduino controlling two I2C devices*

# I2C Hardware

Electrically, I2C interfaces connection lines from a microcontroller or peripheral can act as both a digital output or digital input (also called *tri-state*). In tri-state mode, the connection lines are neither HIGH nor LOW, but are, instead, a floating value. The outputs are also *open-collector*, which means that they require a pull-up resistor. These resistors should be 4.7 kΩ in value, and there should be just one pair for the whole I2C bus, pulling up to either 3.3V or 5V, depending on the voltage at which you want the bus to operate. If some devices on the bus use different voltages, you need to use a level converter. Bidirectional level converter modules suitable for I2C are available, such as the BSS138 device from Adafruit: www.adafruit.com/products/757.

The various Arduino boards allocate different pins to I2C. For example, the Uno uses pins A4 and A5 as SDA and SCL, respectively, whereas the Leonardo uses pins D2 and D3. (More on SDA and SCL in the next section.) On both boards, the SDA and SCL pins are available on the socket header next to the AREF connection (Figure 7-3).

**Figure 7-3**  *I2C connections on an Arduino Uno*

| Board | Pins | Notes |
|---|---|---|
| Uno | A4 (SDA) and A5 (SCL) | The connections labeled SCL and SDA near AREF also connect to A4 and A5. |
| Leonardo | D2 (SDA) and D3 (SCL) | The connections labeled SCL and SDA near AREF also connect to D2 and D3. |
| Mega2560 | D20 (SDA) and D21 (SCL) | |
| Due | D20 (SDA) and D21 (SCL) | The Due also has a second I2C pair of connections that are labeled SDA1 and SCL1 |

**Table 7-1**  *I2C Connections on Arduino Boards*

Table 7-1 indicates the location of I2C pins on the common Arduino boards.

# The I2C Protocol

I2C uses two wires to transmit and receive data (hence, the alternative name of Two Wire Interface). These two lines are called the Serial Clock Line (SCL) and the Serial Data Line (SDA). Figure 7-4 shows the timing diagram for this signal.

The master supplies the SCL clock, and when there is data to be transmitted, the sender (master or slave) takes the SDA line out of tri-state (digital input mode) and sends data as logic highs or lows in time with the clock signal. When transmission is complete, the clock can stop and the SDA pin is returned to tri-state.

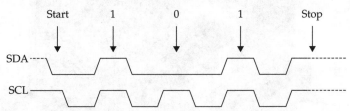

**Figure 7-4**  *Timing diagram for I2C*

# The Wire Library

You could, of course, generate these pulses yourself by *bit banging*—that is, turning digital outputs on and off in your code. To make life easier for us, however, the Arduino software includes a library called Wire that handles all the timing complexity, so we can just send and receive bytes of data.

To use the Wire library, you first need to include it using the following command:

```
#include <Wire.h>
```

## Initializing I2C

In most situations, an Arduino is the "master" in any I2C bus. To initialize an Arduino as the master, use the **begin** command in your **setup** function, as shown here:

```
void setup()
{
  Wire.begin();
}
```

Note that because the Arduino is the master in this arrangement, you don't need to specify an address. If the Arduino were being initialized as a slave, then you would need to specify an address, 0 to 127, as its parameter to uniquely identify it on the I2C bus.

## Master Sending Data

To send data to an I2C device, start by using the **beginTransmission** function and specifying the address of the I2C device on the bus that you wish to send data to:

```
Wire.beginTransmission(4);
```

You can either send data to an I2C device one byte at a time, or you can send a **char** array, as shown in these two examples:

```
Wire.send(123) // send the byte 123
Wire.send("ABC"); // send the string of chars "ABC"
```

Finally, at the end of the transmission, use the **endTransmission** function:

```
Wire.endTransmission();
```

## Master Receiving Data

For a master to receive data from a slave, it must first request the number of bytes it requires using the **requestFrom** function:

```
Wire.requestFrom(4, 6);     // request 6 bytes from slave address 4
```

The first argument to this function is the address of the slave from which the master wants to receive data, and the second argument is the number of bytes that the master is expecting to receive back. The slave can return less than this, so the **available** function is used to determine both if data has arrived and the number of bytes received. The following example (taken from the Wire example sketches) shows the master reading all available data from the slave and echoing it to the Serial Monitor:

```
#include <Wire.h>

void setup()
{
  Wire.begin();          // join i2c bus (address optional for master)
  Serial.begin(9600);    // start serial for output
}

void loop()
{
  Wire.requestFrom(4, 6);     // request 6 bytes from slave address 4

  while(Wire.available())     // slave may send less than requested
  {
    char c = Wire.receive();  // receive a byte as character
    Serial.print(c);          // print the character
  }

  delay(500);
}
```

The Wire library will buffer incoming I2C data.

# I2C Examples

Any I2C device should have an accompanying datasheet that specifies the messages that it expects to use. Sometimes you will need to use that datasheet to build your own messages to send from the Arduino and to interpret the messages that come back. You'll often find, however, that when an I2C device is commonly used with an Arduino, then someone has written a library that wraps the I2C messages in nice easy-to-use functions. In fact, if there isn't a library and you work out how to use the device, then the socially minded thing to do is to release your library to the world and earn yourself some open source karma.

Even if no fully fledged library is available, you can often find useful code snippets for the device on the Internet.

## TEA5767 FM Radio

The first I2C example does not use a library. It deals with raw messages to interface an Arduino with a TEA5767 module. These modules are available at a very low cost on the Internet and are easy to connect to an Arduino to use as an Arduino-controlled FM receiver.

The tricky part is that the connections on these devices are set at an extremely fine pitch, so you generally need to make or buy some kind of adapter that allows you to use them with breadboard or jumper wires.

Figure 7-5 shows how this module can be wired to an Arduino.

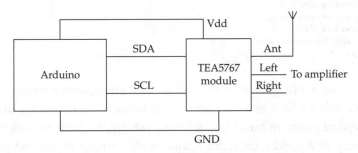

**Figure 7-5**   *Wiring a TEA5767 module to an Arduino Uno using I2C*

You can find the full datasheet for the TEA5767 here: www.sparkfun .com/datasheets/Wireless/General/TEA5767.pdf. The datasheet contains a lot of technical information about the chip, but if you scroll through the document, you'll find a section detailing the messages that it expects to receive. The datasheet specifies that the TEA5767 expects to receive messages of five bytes. The example code shown next is a fully working example that will tune the frequency once at startup. In practice, you need some other mechanism, such as push buttons and an LCD display, to set the frequency.

```
// sketch_07_01_I2C_TEA5767

#include <Wire.h>

void setup()
{
  Wire.begin();
  setFrequency(93.0); // MHz
}

void loop()
{
}

void setFrequency(float frequency)
{
  unsigned int frequencyB = 4 * (frequency * 1000000 + 225000) / 32768;
  byte frequencyH = frequencyB >> 8;
  byte frequencyL = frequencyB & 0XFF;

  Wire.beginTransmission(0x60);
  Wire.write(frequencyH);
  Wire.write(frequencyL);
  Wire.write(0xB0);
  Wire.write(0x10);
  Wire.write(0x00);
  Wire.endTransmission();
  delay(100);
}
```

The code we're interested in is all in the **setFrequency** function. This function takes a float as a parameter. This value is the frequency in MHz. So if you're going to build this for real, you might want to look up the frequency of a good local radio station with a strong signal and put the value in the call to **setFrequency** in the **setup** function.

To convert a float frequency in MHz into a two-byte value that can be sent as part of the five-byte message, you need to do some math. The math is contained in the code:

```
unsigned int frequencyB = 4 * (frequency * 1000000 + 225000) / 32768;
byte frequencyH = frequencyB >> 8;
byte frequencyL = frequencyB & 0XFF;
```

The **>>** command shifts bits to the right, so using **>> 8** shifts the most significant 8 bits into the least significant 8 bit positions. The **&** operator provides a bitwise **and** operation, which has the effect of masking off the top 8 bits so only the bottom 8 bits remain. For more information on this kind of bit manipulation, see Chapter 9.

The remainder of the **setFrequency** function begins transmission of the I2C message to the slave with address **0x60**, which is fixed for the TEA5767 chip. It then sends each of the 5 bytes, starting with the 2 frequency bytes.

If you read through the datasheet, you'll discover many other things you can accomplish with different messages, such as scanning, muting one or more channels, and setting the mode to mono or stereo.

In the Appendix, we'll revisit this example, creating an Arduino library so using the TEA5767 can be even simpler.

# Arduino-to-Arduino Communication

This second example uses two Arduinos, one acting as the I2C master and one as the slave. The master will send messages to the slave, which will, in turn, echo them to the Serial Monitor, so we can see that the communication is working.

The connections for this setup are shown in Figure 7-6. Note that the TEA5767 module has built-in I2C pull-up resistors, but this is not the case when connecting two Arduinos, so you'll need to provide your own 4.7 kΩ resistors, as shown in Figure 7-6.

We need to program each of the two Arduinos with a different sketch. Both sketches are provided as examples in the Wire library. Program the master Arduino with File | Example | Wire | master_writer, and the slave Arduino with File | Example | Wire | slave_receiver.

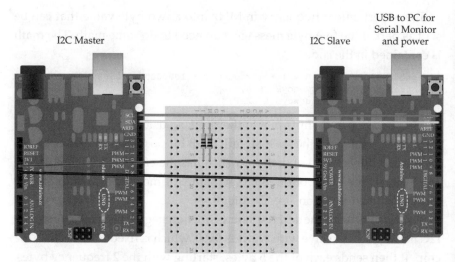

**Figure 7-6**   *Connecting two Arduinos using I2C*

Once you've programmed both Arduinos, leave the slave Arduino connected to your computer; you need to see the output from this Arduino in
the Serial Monitor, and it will also supply power to the master Arduino.

Start with the sketch on the master Arduino:

```
#include <Wire.h>

void setup()
{
  Wire.begin(); // join i2c bus (address optional for master)
}

byte x = 0;

void loop()
{
  Wire.beginTransmission(4);   // transmit to device #4
  Wire.write("x is ");         // sends five bytes
  Wire.write(x);               // sends one byte
  Wire.endTransmission();      // stop transmitting
  x++;
  delay(500);
}
```

This code generates messages of the form *"x is 1"* where *1* is a number that is incremented every half second. This message is then sent to the I2C slave device with the ID of 4, as specified in **beginTransmission**.

The slave sketch's job is to receive the messages coming from the master and echo them on the Serial Monitor:

```
#include <Wire.h>

void setup()
{
  Wire.begin(4);               // join i2c bus with address #4
  Wire.onReceive(receiveEvent); // register event
  Serial.begin(9600);          // start serial for output
}

void loop()
{
  delay(100);
}

// function that executes whenever data is received from master
// this function is registered as an event, see setup()
void receiveEvent(int howMany)
{
  while(1 < Wire.available()) // loop through all but the last
  {
    char c = Wire.read();    // receive byte as a character
    Serial.print(c);         // print the character
  }
  int x = Wire.read();       // receive byte as an integer
  Serial.println(x);         // print the integer
}
```

The first thing to notice is that this time the **Wire.begin** function has a parameter of **4**. This parameter specifies the I2C address of the slave, which is 4. It must match the address that the master sends the message to.

**TIP**   *You could connect many slave Arduinos to the same two-wire bus as long as each has a different I2C address.*

The sketch for the slave differs from that of the master because it uses interrupts to respond to the master when a message comes in. This is accomplished using the **onReceive** function, which is invoked like an interrupt service routine (see Chapter 3). Place this in **setup** so the user-written function **receiveEvent** is invoked whenever a message comes in.

The **receiveEvent** function is expected to have a single parameter, which indicates the number of bytes ready to be read. In this case, this number is ignored. The **while** loop first reads all the available characters and echoes each character in turn. It then reads the single byte number on the end of the message and prints that to the Serial Monitor. Using **println** rather than **write** ensures that the value of the byte is displayed rather than its character value (Figure 7-7).

## LED Backpack Boards

Another common range of I2C devices are those used for displays. Of these, the range of backpack boards for matrix and seven-segment LED displays from Adafruit are typical. They contain an LED display mounted on a circuit board that also has an I2C LED controller chip on it. This setup reduces the normally large number of Arduino I/O pins required for controlling an LED display with just the two I2C SDA and SCL pins.

These devices (top row of Figure 7-1) are used with a pair of libraries that provide a comprehensive set of functions for displaying graphics and text on Adafruit's range of LED backpacks. You can find out more about these colorful and interesting devices here: www.adafruit.com/products/902.

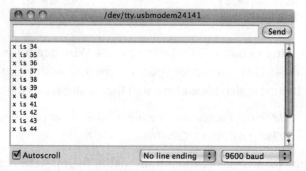

**Figure 7-7**  *Serial Monitor output for Arduino to Arduino over I2C*

Once you've installed the libraries, all the I2C communication is hidden away, and you can just use high-level commands as illustrated by the following code taken from the libraries' example sketches:

```
#include <Wire.h>
#include "Adafruit_LEDBackpack.h"
#include "Adafruit_GFX.h"
Adafruit_8x8matrix matrix = Adafruit_8x8matrix();

void setup()
{
  matrix.begin(0x70);
  matrix.clear();
  matrix.drawLine(0,0, 7,7, LED_RED);
  matrix.writeDisplay();
}
```

# DS1307 Real-Time Clock

Another common I2C device is the DS1307 RTC chip. This chip also has a well-used and reliable Arduino library to simplify it and hide the actual I2C messages. The library is called *RTClib* and can be downloaded from here: https://github.com/adafruit/RTClib.

The fragments of code are, again, taken from the examples supplied with the library.

```
#include <Wire.h>
#include "RTClib.h"

RTC_DS1307 RTC;

void setup () {
    Serial.begin(9600);
    Wire.begin();
    RTC.begin();

  if (! RTC.isrunning()) {
    Serial.println("RTC is NOT running!");
    // sets the RTC to the date & time this sketch was compiled
    RTC.adjust(DateTime(__DATE__, __TIME__));
  }
}
```

```
void loop () {
    DateTime now = RTC.now();
    Serial.print(now.year(), DEC);
    Serial.print('/');
    Serial.print(now.month(), DEC);
    Serial.print('/');
    Serial.print(now.day(), DEC);
    Serial.print(' ');
    Serial.print(now.hour(), DEC);
    Serial.print(':');
    Serial.print(now.minute(), DEC);
    Serial.print(':');
    Serial.print(now.second(), DEC);
    Serial.println();
    delay(1000);
}
```

If you want to see the actual I2C code, then you can open the library files and look at how they work. For example, you'll find the RTClib library in the files **RTClib.h** and **RTClib.cpp**. These files are in the folder **libraries/RTClib**.

For example, you can find the function definition for **now** in **RTClib .cpp**:

```
DateTime RTC_DS1307::now() {
    Wire.beginTransmission(DS1307_ADDRESS);
    Wire.write(i);
    Wire.endTransmission();

    Wire.requestFrom(DS1307_ADDRESS, 7);
    uint8_t ss = bcd2bin(Wire.read() & 0x7F);
    uint8_t mm = bcd2bin(Wire.read());
    uint8_t hh = bcd2bin(Wire.read());
    Wire.read();
    uint8_t d = bcd2bin(Wire.read());
    uint8_t m = bcd2bin(Wire.read());
    uint16_t y = bcd2bin(Wire.read()) + 2000;

    return DateTime (y, m, d, hh, mm, ss);
}
```

The values that are read over I2C are in binary-coded decimal (BCD), which must be converted into bytes using the **bcd2bin** function in the library.

BCD splits a byte into two 4-bit nibbles (yes, really). Each nibble represents one digit of a two-digit decimal number. So the number 37 is represented in a BCD byte as 0011 0111. The first four bits being decimal 3 and the second four bits 7.

# Summary

In this chapter, you have learned about I2C and how to use it with an Arduino to communicate with peripherals and other Arduinos.

In the next chapter, we examine another type of serial bus interface that is used to communicate with peripherals. This interface, called *1-wire*, is not as widely used as I2C, but is used in the popular DS18B20 temperature sensor.

# 8

# Interfacing with 1-Wire Devices

1-Wire is a bus standard designed to serve a similar purpose to the I2C bus (see Chapter 7)—that is, to allow microcontrollers to communicate with peripheral ICs with a minimal number of data lines. The 1-Wire standard created by Dallas Semiconductor has taken this to its logical extreme by reducing the data lines used to just one. The bus is slower than I2C, and it has the interesting feature of *parasitic power*, which allows remote devices to be connected to a microcontroller with just two wires, GND (ground), and combined power and data wire.

The 1-Wire bus standard has a much smaller range of potential devices than I2C, most manufactured by Dallas Semiconductor or Maxim. They include printer cartridge identity devices, EEPROM flash memory, and analog-to-digital converters (ADCs). However, the most commonly used 1-Wire device for hobbyists is the Dallas Semiconductor DS18B20 temperature sensor.

## 1-Wire Hardware

Figure 8-1 shows how you can connect a DS18B20 to an Arduino using just two connections and the DS18B20's parasitic power mode.

1-Wire is a bus, rather than a point-to-point connection, and you can chain together up to 255 devices using the arrangement shown in Figure 8-1.

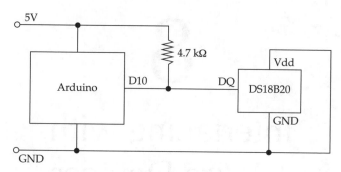

**Figure 8-1**   *Connecting a 1-Wire device to an Arduino*

If you wish to use the device in "normal" power mode, then you can omit the 4.7 kΩ resistor and connect Vdd on the DS18B20 directly to 5V from the Arduino instead of to GND.

# The 1-Wire Protocol

Just as with I2C, 1-Wire uses the master and slave concept for devices. The microcontroller is the master and the peripherals are the slaves. Each slave device is given a unique ID known as its "address" during manufacturing, so it can be identified on the bus when there are multiple slaves. This address is 64 bits in length, allowing for roughly $1.8 \times 10^{19}$ different IDs.

The protocol is similar to I2C in that the bus line is switched from being an input to being an output by the master to allow two-way communication. However, rather than have separate clock and data signals, 1-Wire has just a single data line and uses long and short pulses to signify 1s and 0s. A pulse of 60 µS signifies a 0 and 15 µS indicates a 1.

The data line is normally HIGH, but when the microcontroller (master) needs to send a command to the device, it sends a special "reset" LOW pulse of at least 480 microseconds. The stream of 1 and 0 pulses then follow this.

# The OneWire Library

The use of 1-Wire is greatly simplified by the OneWire library, which you can download from http://playground.arduino.cc/Learning/OneWire.

# Initializing 1-Wire

The first step in using an Arduino as the master on a 1-Wire bus is to import the OneWire library using this command:

```
#include <OneWire.h>
```

The next step is to create an instance of OneWire and specify the Arduino pin to be used for the 1-Wire data bus. You can combine these into a single command, and you can use any Arduino pin for the bus; simply supply the pin number as the parameter:

```
OneWire bus(10);
```

In the example, the bus will be initialized on pin D10 of the Arduino.

# Scanning the Bus

Because each slave device on the bus is allocated a unique ID during manufacturing, you need a way to find the devices on the network. It would be unwise to hard-code device addresses into the Arduino sketch because if you were to replace one of the slave devices, the new device would have a different address than the old one and you wouldn't be able to use it. So the master (Arduino) can essentially produce a list of the devices on the bus. What is more, the first 8 bits of the address indicate the "family" of device, so you can tell if the device is, say, a DS18B20 or some other type of device.

Table 8-1 lists some of the most common family codes for 1-Wire. You can find a more complete list here: http://owfs.sourceforge.net/family.html.

| Family Code (Hexadecimal) | Device Family | Description |
|---|---|---|
| 06 | iButton 1993 | Identity button chip |
| 10 | DS18S20 | Precision temperature sensor (9-bit resolution) |
| 28 | DS18B20 | Precision temperature sensor (12-bit resolution) |
| 1C | DS28E04-100 | 4KB EEPROM |

**Table 8-1**  *Family Codes for 1-Wire Addresses*

The OneWire library has a **search** function that you can use to find all the slave devices on the bus. The following example code lists the addresses of all the devices on the bus to the Serial Monitor:

```
// sketch_08_01_OneWire_List

#include <OneWire.h>

OneWire bus(10);

void setup()
{
  Serial.begin(9600);
  byte address[8]; // 64 bits
  while (bus.search(address))
  {
    for(int i = 0; i < 7; i++)
    {
      Serial.print(address[i], HEX);
      Serial.print(" ");
    }
    // checksum OK or Fail
    if (OneWire::crc8(address, 7) == address[7])
    {
      Serial.println(" CRC OK");
    }
    else
    {
      Serial.println(" CRC FAIL");
    }
  }
}

void loop()
{
}
```

Figure 8-2 shows the result of running this sketch with two DS18B20 temperature sensors attached to an Arduino. Note that for both devices, the "family" code is contained in the first byte and is 28 (hexadecimal) in both cases.

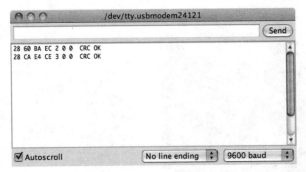

**Figure 8-2**  *Listing 1-Wire slave devices*

The **search** function requires an array of 8 bytes in which to put the next address that it finds. If no more devices are to be found, it will return 0. This allows the **while** loop in the previous example to keep iterating until all the addresses have been displayed. The last byte of the address is actually a cyclic redundancy check (CRC) that ensures the integrity of the address. The OneWire library includes a CRC checking function.

# Using the DS18B20

The following example illustrates the use of OneWire with the DS18B20 temperature sensor. Figure 8-3 shows a DS18B20 chip connected to an Arduino. Note how the chip itself is just a three pin transistor-like chip.

The Dallas Semiconductor temperature sensor has its own library that makes requesting the temperature and decoding the result easier. The DallasTemperature library can be downloaded from here: https://github .com/milesburton/Arduino-Temperature-Control-Library.

```
// sketch_08_02_OneWire_DS18B20

#include <OneWire.h>
#include <DallasTemperature.h>

const int busPin = 10;

OneWire bus(busPin);
DallasTemperature sensors(&bus);
DeviceAddress sensor;
```

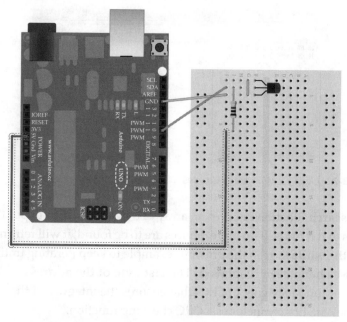

**Figure 8-3** *A DS18B20s connected to an Arduino*

```
void setup()
{
  Serial.begin(9600);
  sensors.begin();
  if (!sensors.getAddress(sensor, 0))
  {
    Serial.println("NO DS18B20 FOUND!");
  }
}

void loop()
{
  sensors.requestTemperatures();
  float tempC = sensors.getTempC(sensor);
  Serial.println(tempC);
  delay(1000);
}
```

This example displays the temperature in Celsius from a single temperature sensor in the Serial Monitor window (Figure 8-4).

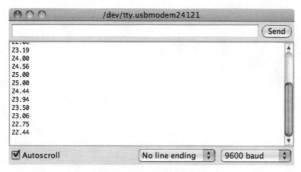

**Figure 8-4**   *Displaying the temperature using a DS18B20*

This example uses just one temperature sensor, but you can easily extend it to use multiple sensors. The DallasTemperature library wraps the OneWire address discovery process in the **getAddress** function, the second parameter of which is the index position of the sensor. To add a second sensor, you need to add a new variable for its address and then set that address using **getAddress**. You can download an example of using two sensors from the book's website as **sketch_08_03_OneWire_DS18B20_2**.

# Summary

In this chapter, you learned a little about the 1-Wire bus and how to use it with the popular DS18B20 temperature sensor.

In the next chapter, we look at yet another kind of serial data interface called SPI.

# 9

# Interfacing with SPI Devices

The Serial Peripheral Interface (SPI) bus is yet another serial bus standard that you can use to connect peripherals to your Arduino. It is fast but uses four pins compared with the two that I2C uses. SPI is not actually a true bus, as the fourth pin is a Save Select (SS) pin. One Arduino pin must be used for SS for each peripheral on the bus. This setup effectively addresses the right peripheral on the bus by turning all the other peripherals off.

A wide range of SPI devices are available, including many of the same type of devices available for I2C. It is not uncommon for peripherals to have both I2C and SPI interfaces.

## Bit Manipulation

SPI interfacing tends to involve a lot of bit manipulation to get data on and off the bus. The first example project (using an MCP3008 ADC IC), in particular, requires a good understanding of how to shuffle bits along and mask the ones you don't want in order to extract an integer value for the analog reading. For this reason, before I go any further into the workings of SPI, I'll make a diversion to explain, in more detail, bit manipulation.

## Binary and Hex

You first met the concept of bits and bytes back in Chapter 4 (see Figure 4-2). When you are manipulating bits in a byte or word (two bytes), you can use their decimal values, but converting between binary and decimal is not that easy to do in your head. For this reason, values are often expressed as binary constants in Arduino C, which you can do using the special syntax shown in this example:

```
byte x = 0b00000011; // 3
unsigned int y = 0b0000000000000011; // 3
```

In the first line, a byte with the decimal value of 3 (2 + 1) is defined. The leading zeros are optional, but providing them serves as a handy reminder that 8 bits are available.

The second example uses an **int** to hold 16 bits. The qualifier **unsigned** is placed in front of **int** to indicate that the variable should only be used to represent positive numbers. This qualifier only really matters if you are using +, −, *, and so on, with the variable, which you should not do if you are using it for bit manipulation. But including the word **unsigned** is good practice.

When you get to 16 bits, the binary representation starts to look a bit long and unwieldy. For this reason, people often use a notation called *hexadecimal*, or more commonly just *hex*, to represent longer binary numbers.

Hex is number base 16, which means you have the usual digits 0 to 9 but also the letters A to F that represent the decimal values 10 to 15. That way, each four bits of a number can be represented in a single digit. Table 9-1 shows the decimal, binary, and hexadecimal representations of the numbers 0 to 15 (decimal).

Hex constants have a special notation similar to that of binary:

```
int x = 0x0003; //   3
int y = 0x010F; // 271 (256 + 15)
```

You'll see this notation used outside of C, in documentation, to make it clear that the number is hex and not decimal.

| Decimal | Binary | Hex |
|---|---|---|
| 0 | 0000 | 0 |
| 1 | 0001 | 1 |
| 2 | 0010 | 2 |
| 3 | 0011 | 3 |
| 4 | 0100 | 4 |
| 5 | 0101 | 5 |
| 6 | 0110 | 6 |
| 7 | 0111 | 7 |
| 8 | 1000 | 8 |
| 9 | 1001 | 9 |
| 10 | 1010 | A |
| 11 | 1011 | B |
| 12 | 1100 | C |
| 13 | 1101 | D |
| 14 | 1110 | E |
| 15 | 1111 | F |

**Table 9-1**   *Binary and Hexadecimal Numbers*

# Masking Bits

A common problem when you receive data from a peripheral using any kind of connection is that the data arrives packed into bytes and not all of the bytes are needed. Peripheral designers often fit as much information as they can into as few bits as possible, speeding up communication, but often at the expense of making the devices more difficult to program.

The process of "masking" bits allows you to disregard some of the data in a byte or larger data structure. Figure 9-1 shows how a byte containing multiple data can be masked to produce a number from the least significant three bits of the byte.

You'll come across the phrases "least significant" and "most significant" to describe binary numbers. In binary written in the normal mathematical way, the most significant bit is the leftmost bit and the least significant bit is the rightmost. After all, the rightmost is only worth 1 or 0. You'll also see the terms *most significant bit (MSB)* and *least significant bit*

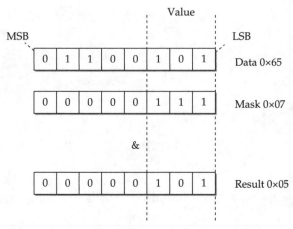

**Figure 9-1**  *Masking bits*

*(LSB).* The least significant bit is also sometimes referred to as *bit 0,* bit 1 being the next most significant bit and so on.

In the example shown in Figure 9-1, the data byte has some values at the most significant end that we are not interested in and only three bits at the least significant end that we want to extract as a number. You do this by "anding" the data with a mask value that has the three bits you're interested in set to 1. Then you "and" together two bytes; each of the bits is, in turn, "anded" with each other to build a result. The result of "anding" two bits is only 1, if both the bits are 1.

Here's how the example looks in Arduino C using the **&** operator. Note that bitwise **and** uses the single **&** character rather than the **&&** used in logical **and**.

```
byte data = 0b01100101;
byte result = (data & 0b00000111);
```

At the end, the variable "result" contains the value 5 (decimal).

## Shifting Bits

Another thing you will find with received data is that having masked the bits you want, those bits are not all at the least significant end of the byte.

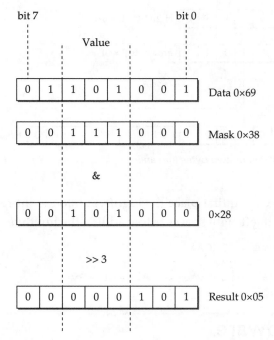

**Figure 9-2**  *Masking and shifting bits*

For example, if the value of interest in the data used in Figure 9-1 was between bits 5 and 3 (see Figure 9-2), you need to first mask the bits of interest, as you did in the previous example, and then sift the bits three places to the right.

You use the C operator >> to shift bits to the right and the number following the >> is the number of bit positions to shift the bits. This may result in some bits being shifted off the end of the byte. Here's this example written in C:

```
byte data = 0b01101001;
byte result = (data & 0b00111000) >> 3;
```

What if you need to take two 8-bit bytes and assemble them into a single 16-bit **int**? You can accomplish this by first shifting the bits of one byte (the most significant byte) to one end of the **int** and then adding in the second byte. Figure 9-3 illustrates this process.

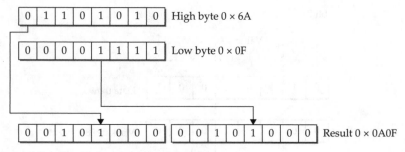

**Figure 9-3** *Combining two bytes into an int*

In Arduino C, you first place the **highByte** into the **int** result variable and then shift it left eight spaces before adding the **lowByte**:

```
byte highByte = 0x6A;
byte lowByte = 0x0F;
int result = (highByte << 8) + lowByte;
```

# SPI Hardware

Figure 9-4 shows a typical configuration for an Arduino with two slave devices.

On the Arduino, the System Clock (SCLK), Master Out Slave In (MOSI), and Master In Slave Out (MISO) are linked to the Arduino pins of the

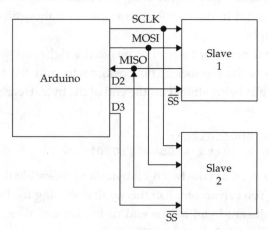

**Figure 9-4** *Arduino and two slave SPI devices*

| Board | SCLK | MOSI | MISO |
|---|---|---|---|
| Uno | 13 (ICSP3) | 11 (ICSP4) | 12 (ICSP1) |
| Leonardo | ICSP3 | ICSP4 | ICSP1 |
| Mega2560 | 52 (ICSP3) | 51 (ICSP4) | 50 (ICSP1) |
| Due | ICSP3 | ICSP4 | ICSP1 |

**Table 9-2**    *SPI Connections on Arduino Boards*

same name, which map to pins D13, D11, and D12 on an Arduino Uno. Table 9-2 lists the pin assignments on the most common Arduino boards.

The Slave select pins can be any pins on the Arduino. They are used to enable a particular slave just before data transmission and then disable it after communication is complete.

No pull-up resistors are required on any of the lines.

Because some Arduino boards, including the Leonardo, only have SPI connectors that are accessible from the ICSP header pins, shields that use SPI often have a socket header that meets the ICSP male header. Figure 9-5 shows the ICSP header with the ICSP headers labeled.

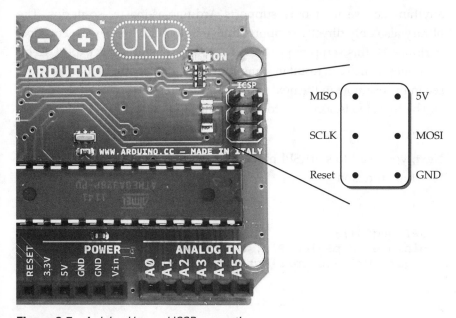

**Figure 9-5**    *Arduino Uno and ICSP connections*

Note that the Arduino Uno has a second ICSP header near the reset button. This is for programming the USB interface.

# The SPI Protocol

The SPI protocol is, at first sight, confusing because data is transmitted and received at the same time by both the master and the currently selected slave. At the same time that the master (Arduino) sends a bit from its MOSI pin to the corresponding MOSI pin on the slave another bit is being sent back from the Slave's MISO pin to the Arduino's MISO pin.

Typically, the Arduino sends a byte's worth of bits and then sends eight zeros while, at the same time, reading the results coming back from the slave. Because the master sets the transmission frequency, make sure the rate is not too fast for the slave device.

# The SPI Library

The SPI library is included with Arduino IDE, so you do not need to install anything to use it. It only supports Arduino-as-master scenarios. The library also only directly supports transmission of whole bytes. For most peripherals, this setup is just fine; however, some devices expect 12-bit messages, which can result in some complicated bit manipulation as you'll see in the example in the next section of this chapter.

The first step is, as usual, to include the SPI library:

```
#include <SPI.h>
```

Next, you need to start SPI by issuing the **SPI.begin** command in your "startup" function.

```
void setup()
{
  SPI.begin();
  pinMode(chipSelectPin, OUTPUT);
  digitalWrite(chipSelectPin, HIGH);
}
```

Unless you are using a Due, you also need to set up digital outputs for each of the SS pins to the slave devices. These outputs can be any Arduino pins. Having set them to be outputs, you need to set them to HIGH immediately because the slave select logic is inverted, so a LOW means the slave is selected.

The Due has extended the SPI library so you can specify one pin to be used for slave selecting, and then the library automatically sets this LOW before transmission and then HIGH after transmission is complete. You can use this feature simply by specifying the pin to use as the only argument to **SPI.begin**. The disadvantage of doing it this way, however, is that it breaks compatibility with other Arduino boards. In the examples in this chapter, all the slave select pins are controlled manually and are, therefore, suitable for all Arduino boards.

A number of utility functions allow you to configure the SPI connection. However, the defaults will normally work, so you only need to change these settings if the datasheet for the slave device leads you to believe they might need changing. These functions are summarized in Table 9-3.

The combined data send and receive happens in the **transfer** function. This function transfers a byte of data and returns the byte of data that it received during the send operation.

```
byte sendByte = 0x23;
byte receiveByte = SPI.transfer(sendByte);
```

| Function | Description |
|---|---|
| SPI.setClockDivider(SPI_CLOCK_DIV64) | Divide the default clock frequency of 4 MHz by 2, 4, 8, 16, 32, 64, or 128. |
| SPI.setBitOrder(LSBFIRST) | Set the bit order to **LSBFIRST** or **MSBFIRST**. The default is **MSBFIRST**. |
| SPI.setDataMode(SPI_MODE0) | The possible values for this function are **SPI_MODE0** to 3. This determines the polarity and phase of the clock signal. Under normal circumstances, you do not need to change this, unless the datasheet indicates a particular mode for the device. |

**Table 9-3**  *Configuration Functions*

Because a conversation with a peripheral usually takes the form of the master requesting something from the slave and the slave responding, you'll often have two transfers in order: one to request the data and the other (a send, probably of 0s) to pull back the data from the peripheral. You'll see this in the next example.

# SPI Example

This example interfaces a MCP3008 eight-channel ADC IC to an Arduino, adding another eight 10-bit analog inputs to your Arduino. The chip is low cost and easy to wire.

Figure 9-6 shows the chip wired to the Arduino using breadboard and jumper wires. The variable resistor (pot) is used to vary the voltage to analog input 0 between 0 and 5V.

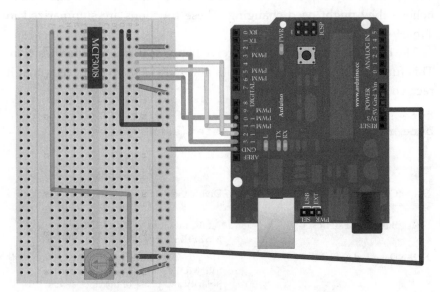

**Figure 9-6** *Wiring diagram for SPI example*

Following is the sketch for this example:

```
// sketch_09_01_SPI_ADC

#include <SPI.h>

const int chipSelectPin = 10;

void setup()
{
  Serial.begin(9600);
  SPI.begin();
  pinMode(chipSelectPin, OUTPUT);
  digitalWrite(chipSelectPin, HIGH);
}

void loop()
{
  int reading = readADC(0);
  Serial.println(reading);
  delay(1000);
}

int readADC(byte channel)
{
  unsigned int configWord = 0b11000 | channel;
  byte configByteA = (configWord >> 1);
  byte configByteB = ((configWord & 1) << 7);
  digitalWrite(chipSelectPin, LOW);
  SPI.transfer(configByteA);
  byte readingH = SPI.transfer(configByteB);
  byte readingL = SPI.transfer(0);
  digitalWrite(chipSelectPin, HIGH);

//  printByte(readingH);
//  printByte(readingL);

  int reading = ((readingH & 0b00011111) << 5) + ((readingL & 0b11111000) >> 3);

  return reading;
}

void printByte(byte b)
{
  for (int i = 7; i >= 0; i--)
  {
    Serial.print(bitRead(b, i));
  }
  Serial.print(" ");
}
```

The function **printByte** was just used during development to display the binary data. Although **Serial.print** can display binary values, it does not include leading zeros, which makes interpreting the data difficult, whereas the **printByte** function always prints all 8 bits.

To see the data coming from the MCP3008, you can remove the // before the two calls to **printByte** and the binary data you are interested in will be displayed.

All the interesting code happens in the **readADC** function, which takes the ADC channel (0 to 7) as its parameter. The first thing you need to do is to use some bit manipulation to create the configuration byte that specifies the kind of analog conversion you want to perform and also the channel you want to use.

The chip is capable of two ADC operation modes. One mode is to compare two analog channels, and the second mode (which this example uses) returns the single-ended reading from the channel specified, just like an Arduino analog input. The datasheet for the MCP3008 (http://ww1.microchip.com/downloads/en/DeviceDoc/21295d.pdf) specifies that the configuration command needs to set four bits: the first bit needs to be 1 for single-ended mode; the next three bits determine the channel (0 to 7) to use.

The MCP3008 is not designed for the byte-at-a-time way in which the SPI library works. In order for the MCP3008 to recognize these 4 bits, we have to split them across 2 bytes. Here's the code for doing this:

```
unsigned int configWord = 0b11000 | channel;
byte configByteA = (configWord >> 1);
byte configByteB = ((configWord & 1) << 7);
```

The first byte of the configuration message contains two 1s, the first of which may not be needed and the second 1 corresponding to the mode bit (single-ended). The other 2 bits in this byte are the most significant 2 bits of the analog channel number. The remaining bit of this number is in the second configuration byte as its most significant bit.

The next line sets the SS line for the chip **LOW** to enable it.

```
digitalWrite(chipSelectPin, LOW);
```

After that, the first configuration byte is sent:

```
SPI.transfer(configByteA);
  byte readingH = SPI.transfer(configByteB);
```

```
byte readingL = SPI.transfer(0);
digitalWrite(chipSelectPin, HIGH);
```

The analog data will not start arriving until the second configuration byte is sent. The 10 bits of data from the ADC are split across 2 bytes, so to flush out the remaining data, a call is made to "transfer" sending a byte load of zeros.

The SS output is now set **HIGH** as the communication is now complete.

Finally, the actual 10-bit analog reading value is calculated using the following line:

```
int reading = ((readingH & 0b00011111) << 5)
            + ((readingL & 0b11111000) >> 3);
```

Each of the 2 bytes has 5 of the 10 bytes of data in it. The first byte contains these bits in its least significant 5 bits. All the bits apart from those 5 are masked out and shifted five positions up in the 16-bit **int**. The lower byte contains the remainder of the reading in its most significant five digits. These must be masked and shifted right by three bit positions before they can also be added into the 16-bit **int**.

To test this, open the Serial Monitor. You should see some data appear. If you sweep the slider of the pot clockwise from 0 to 5V, you should see something similar to what's shown in Figure 9-7. The first two binary numbers are the 2 bytes from the MCP3008 and the final decimal number is the analog reading between 0 and 1023.

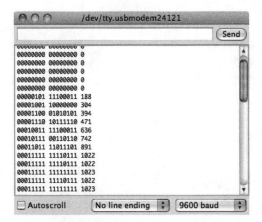

**Figure 9-7**  *Viewing the messages in binary*

# Summary

Interfacing with SPI when no library is available is by no means easy. You will sometimes need to perform a little trial and error to get things going. As with any type of debugging, always start by gathering evidence and examining the data that you are receiving. You will slowly get a picture of what is happening and then be able to tailor your code to produce the desired results.

Text chapter examines the final interface standard supported by the Arduino, that of TTL Serial. This standard is a point-to-point interface rather than a bus, but nonetheless a much-used and handy mechanism for sending and receiving data.

# 10

# Serial UART Programming

You should already be fairly familiar with the serial interface. You use it when you program your Arduino board, and you also use it to communicate with the Serial Monitor to send data back and forth to the Arduino from your computer. You do this through the Arduino's USB-to-serial adapter or directly with the serial adapter. Interfacing directly is often referred to as *TTL Serial*, or just *Serial*. TTL is a reference to Transistor Transistor Logic, a now redundant technology that used 5V logic levels.

Serial communication, of this kind, is not a bus. It is point-to-point communication. Only two devices are involved—generally an Arduino and a peripheral.

Peripherals that use TTL Serial rather than I2C or SPI tend to be larger devices or devices that have been around for a long time and traditionally always had a TTL Serial interface. This also includes devices originally intended to be connected to the serial port of a PC. Examples include GPS modules, multimeters with data logging features, and barcode and RFID readers.

## Serial Hardware

Figure 10-1 shows the serial hardware for the Arduino Uno.

The ATmega328 on the Arduino Uno has two pins Rx and Tx (*Receive* and *Transmit,* respectively). These also double as pins D0 and D1, but if you use them as general I/O pins, you will probably find that you cannot program your Arduino while they are attached to external electronics.

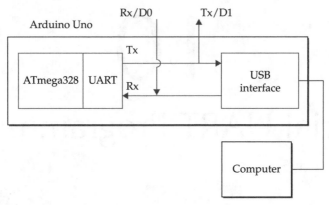

**Figure 10-1** *Arduino Uno serial hardware*

These Rx and Tx pins are the serial interface of the hardware Universal Asynchronous Receiver Transmitter (UART) on the ATmega328. This part of the microcontroller is responsible for sending and receiving bytes of data from and to the microcontroller.

The Uno has a separate processor that acts as a USB-to-serial interface. As well as electrical differences in the serial signal, the USB bus also has a much more complicated protocol than serial and so it does a fair bit of work behind the scenes so it appears the serial port of the ATmega328 is communicating directly with your computer.

The Arduino Leonardo does not have a separate chip to act as an USB interface; rather it uses an ATmega chip that includes two UARTs and a built-in USB interface (Figure 10-2).

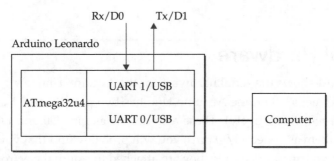

**Figure 10-2** *Arduino Leonardo serial hardware*

| Board | Number of Serial Ports | Details |
|---|---|---|
| Uno | 1 | Rx is D0 and Tx is D1. These ports are also used by USB. |
| Leonardo | 2 | Dedicated USB. Second serial port. Rx is D0 and Tx is D1 |
| Mega2560 | 4 | USB uses D0 and D1. Three other ports: Serial1 on pins 19 (Rx) and 18 (Tx), Serial2 on pins 17 (Rx) and 16 (Tx), Serial3 on pins 15 (Rx) and 14 (Tx). |
| Due | 4 | Dedicated USB. Serial port 0 uses D0 (Rx) and D1 (Tx). Three other ports: Serial1 on pins 19 (Rx) and 18 (Tx), Serial2 on pins 17 (Rx) and 16 (Tx), Serial3 on pins 15 (Rx) and 14 (Tx). |

**Table 10-1**   *UART Serial Interfaces by Arduino Board*

One of the UARTs is dedicated to the USB interface and the other is connected to the Rx and Tx pins (D0 and D1). This gives you the advantage of connecting the Tx and Rx to other electronics and still being able to program the Arduino and send data to the Serial Monitor.

Other Arduino boards have differing quantities and arrangements of serial ports. These are summarized in Table 10-1. Note that the Due is alone among Arduino boards in operating its serial ports at 3.3V rather than 5V.

TTL Serial has a relatively short range (a few feet or perhaps tens of feet), especially if you use it at a high baud rate. For communicating over longer distances, an electrical standard called RS232 has been defined. Until perhaps the last decade, you could commonly find PCs with RS232 serial ports. The RS232 standard changes the signal levels, making them more suitable for traveling a greater distance than with TTL Serial.

# Serial Protocol

The Serial protocol and much of the terminology around it dates back to the early days of computer networking. Both the sender and receiver have to agree on a speed at which to exchange data. This speed, called the *baud rate,* is set at both ends before communication begins. The baud rate is the number of signal transitions per second, which would be the same as the number of bits per second, were it not for the fact that a byte of data may have start, end, and parity bits. So, as a rough approximation, if you divide the baud rate by 10, you'll know about how many bytes per second you can transfer.

Baud rates are selected from a number of standard baud rates. You may have seen these on the Serial Monitor drop-down list on the Arduino IDE. The baud rates used by the Arduino software are: 300, 1200, 4800, 9600, 14400, 19200, 28800, 38400, 57600, and 115200 baud.

The most commonly used baud rate for the Arduino is probably 9600, which tends to be the default baud rate. There is no particularly good reason for this as the Arduino communicates reliably at 115200 baud. For projects that require really fast data transfer, this rate will be used. Another common rate is 2400 baud. Some peripherals such as Bluetooth serial adaptors and GPS hardware use this rate.

Another rather confusing Serial connection parameter that you might encounter is a string of characters like this: 8N1. This string means 8 bits per packet, No parity checking, and 1 stop bit. Although other combinations are possible, any device that you are likely to encounter will be 8N1.

## The Serial Commands

The Serial commands are not contained in a library, so you do not need an **include** command in your sketch.

Start serial communication using the command **Serial.begin**, which takes the baud rate parameter:

```
Serial.begin(9600)
```

This is typically called just once in the **setup** function.

If you are using a board that has more than one serial port, and if you are using the default port (port 0), you just use the **Serial.begin** command. If you are using one of the other ports, however, then put the number after the word **Serial**. For example, to start communication on serial port 3 on an Arduino Due, you would write the following in your sketch:

```
Serial3.begin(9600);
```

Once **Serial.begin** has been called, the UART will listen for incoming bytes and automatically store them in a buffer, so even if the processor is busy doing other things, the bytes will not be lost as long as the buffer does not overflow.

Your **loop** function can check for incoming bytes of data using the **Serial.available** function. This function returns the number of bytes available for reading. If no bytes are available, then it returns 0. This equates to "false" in C, so you will often see code like this that tests for available data:

```
void loop()
{
  if (Serial.available())
  {
    // read and process the next byte
  }
}
```

The **read** command takes no arguments and simply reads the next available byte from the buffer.

The **readBytes** function reads available bytes into a buffer within the sketch, as opposed to the buffer used by the UART. It takes two arguments: the buffer to fill (this should be a reference to an array of bytes) and the maximum number of bytes to read. This argument can be useful if you have a project that needs to send variable length strings to the Arduino. In general, it is better to avoid this, however, and try to make any communication to an Arduino of a fixed length and as simple as possible.

The **parseInt** and **parseFloat** functions can be convenient, as they allow strings sent to the Arduino to be read as numbers into **int** and **float** variables, respectively.

```
void loop()
{
  if (Serial.available())
  {
    int x = parseInt();
  }
}
```

Both functions read characters until they run out or reach a space or other nonnumeric character and then turn the string into a numeric value.

Before using functions like **parseInt** and **parseFloat**, make sure you understand why you are doing this. I have seen code that people have written converting an **int** into an array of characters, that sends the array of characters to a second Arduino, which then turns the array back into an **int**. There are a number of reasons why this is not a good idea:

- It is unnecessary. Serial communication sends binary just fine. All that is required is to send the upper and lower bytes of the **int**, putting them into the upper and lower bytes of a new **int** on receipt.
- Converting numbers into strings and vice versa is slow.
- The serial link may be passing six characters of data (including the null terminator) rather than the 2 bytes of an int.

If the device you are interfacing with is outside of your control and the designer's protocol uses strings to hold numbers, or has variable length fields of data, then these functions can be useful. Otherwise, if the protocol is completely under your control, make life easy for yourself and avoid the unnecessary complexity of converting types and variable-length messages of different formats.

The examples in the "Serial Examples" section, later in this chapter, also serve as templates for designing your own communication code.

Serial has a lot of functions, many of which you'll never need to use. The most handy have been covered here. For the rest, please refer to the Arduino Serial documentation here: http://arduino.cc/en/Reference/Serial.

# The SoftwareSerial Library

Sometimes, especially when using an Arduino Uno, having just one serial port is not enough. The SoftwareSerial library allows you to use almost any pair of pins for serial communication, but with a few limitations:

- You can only receive data from one  SoftwareSerial port at a time.
- You may have trouble using it if your sketch uses timer or external interrupts.

The functions available mirror those of Serial and, in some respects, are better thought out. SoftwareSerial includes support for serial communication for devices that use inverted signals, such as the MaxSonar

| Board | Pins for Tx | Pins for Rx |
|---|---|---|
| Uno | Any, except 0 and 1 | Any, except 0 and 1 |
| Leonardo | Any, except 0 and 1 | 8, 9, 10, 11, 14 (MISO), 15 (SCK), 16 (MOSI) |

**Table 10-2**   *Pin Usage for SoftwareSerial by Arduino Board*

rangefinders. You also create a SoftwareSerial object for each connection, which is cleaner than the standard Arduino approach of putting a number after the word **Serial**.

Table 10-2 shows the pin allocations you can use with SoftwareSerial for the Uno and Leonardo boards. If you are using a bigger board with four hard serial ports, you are unlikely to need SoftwareSerial. Unless prefixed with an *A*, the pin numbers refer to digital pins.

When starting a SoftwareSerial connection, specify the Rx and Tx pins as the two parameters when creating a SoftwareSerial object. Then use **begin** with a baud rate as a parameter to start communication:

```
#include <SoftwareSerial.h>

SoftwareSerial mySerial(10, 11); // RX, TX

void setup()
{
  mySerial.begin(9600);
  mySerial.println("Hello, world?");
}
```

You can find full documentation for the SoftwareSerial library here: http://arduino.cc/en/Reference/SoftwareSerial.

# Serial Examples

This section includes a mix of UART and SoftwareSerial usage examples.

## Computer to Arduino over USB

This first example uses the Serial Monitor to send commands to an Arduino. The Arduino will also send analog readings from A0 once per second, while, at the same time, looking for single-character incoming messages of

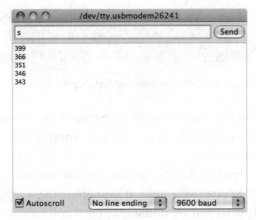

**Figure 10-3**   *The Serial Monitor communicating with Arduino*

*g* for "go" or *s* for "stop" to control the flow of readings. Figure 10-3 shows the Serial Monitor while this sketch is running.

In this situation, because the readings from the Arduino are going to be displayed directly in the Serial Monitor window, the readings may as well be sent as text rather than binary.

Here is the sketch for this example:

```
// sketch_10_01_PC_to_Arduino

const int readingPin = A0;

boolean sendReadings = true;

void setup()
{
  Serial.begin(9600);
}

void loop()
{
  if (Serial.available())
  {
    char ch = Serial.read();
    if (ch == 'g')
    {
      sendReadings = true;
    }
```

```
    else if (ch == 's')
    {
        sendReadings = false;
    }
}
if (sendReadings)
{
    int reading = analogRead(readingPin);
    Serial.println(reading);
    delay(1000);
}
}
```

The **loop** tests for incoming serial data, and if there is any, it reads one byte as a character. This byte is then compared to the **'s'** and **'g'** commands and a status variable, **'sendReadings'**, is set accordingly.

The **sendReadings** variable is then used to determine if the reading should be made and then printed. If the 'sendReadings' flag is true, then there is a second delay before the next reading is sent.

Using **delay** means that **sendReadings** can only be changed the next time around the loop. This is not a problem for this sketch, but in other circumstances you might need a better solution that does not block the loop. See Chapter 14 for more discussion on this kind of thing.

## Arduino to Arduino

This second example illustrates the sending of data from one Arduino Uno to another over a serial connection. In this case, readings from A1 of one Arduino are transmitted to the second Arduino, which then uses them to control the flashing rate of the built-in "L" LED.

The Arduinos are wired together as shown in Figure 10-4.

One Arduino's Tx should be connected to the Rx of the other and vice-versa. In this example, both the Arduinos are using the SoftwareSerial library with pin 8 used as Rx and pin 9 as Tx.

The GND connections of the two Arduinos need to be connected, as do the 5V pins as you want to use the sending Arduino to power the receiving Arduino. The sending Arduino has a trimpot (small variable resistor) pushed into pins A0 to A2. By setting A0 and A2 to be outputs and then setting A2 HIGH, you can vary the voltage at A1 between 0 and 5V by

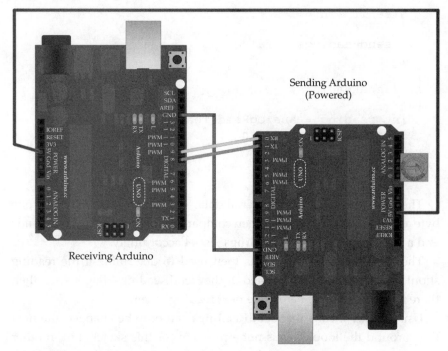

**Figure 10-4** *Two Arduino Unos communicating over serial*

rotating the knob on the trimpot to control the flashing rate of the LED on the other Arduino.

The sending Arduino's sketch is shown here:

```
// sketch_10_02_Adruino_Sender

#include "SoftwareSerial.h"

const int readingPin = A1;
const int plusPin = A2;
const int gndPin = A0;

SoftwareSerial sender(8, 9); // RX, TX

void setup()
{
  pinMode(gndPin, OUTPUT);
  pinMode(plusPin, OUTPUT);
```

```
  digitalWrite(plusPin, HIGH);
  sender.begin(9600);
}

void loop()
{
  int reading = analogRead(readingPin);
  byte h = highByte(reading);
  byte l = lowByte(reading);
  sender.write(h);
  sender.write(l);
  delay(1000);
}
```

To send the 16 bit (**int**) reading, the reading is split into high and low bytes and each byte is then sent over the serial link using **write**. Whereas **print** and **println** convert their argument into a string of characters, **write** sends the byte as binary.

Here is the receiving code:

```
// sketch_10_03_Adruino_Receiver

#include "SoftwareSerial.h"

const int ledPin = 13;
int reading = 0;
SoftwareSerial receiver(8, 9); // RX, TX

void setup()
{
  pinMode(ledPin, OUTPUT);
  receiver.begin(9600);
}

void loop()
{
  if (receiver.available() > 1)
  {
    byte h = receiver.read();
    byte l = receiver.read();
    reading = (h << 8) + l;
  }
  flash(reading);
}
```

```
void flash(int rate)
{
  // 0   slow 1023 very fast
  int period = (50 + (1023 - rate) / 4);
  digitalWrite(ledPin, HIGH);
  delay(period);
  digitalWrite(ledPin, LOW);
  delay(period);
}
```

The receiving code must wait until at least 2 bytes are available and then reconstruct the **int** reading by pushing the high byte up to the top 8 bits of the **int** and then adding the low byte.

If you are considering sending more complex data from one Arduino to another, then you might like to look at the EasyTransfer library: www .billporter.info/2011/05/30/easytransfer-arduino-library/.

Although this example uses wires to connect the Tx for one Arduino to the Rx of another, you could almost accomplish this as easily with wireless connections. Many wireless modules operate transparently, in other words, as if the serial ports were connected by wires.

## GPS Module

The final serial communication example reads positional information (latitude and longitude) from a Global Positioning System (GPS) module using TTL Serial, which then formats the data and sends it to the Serial Monitor (Figure 10-5).

The communication with the GPS module is one way, so only the Tx output of the module needs to be connected to an Rx pin on an Arduino. The module used is a Sparkfun Venus GPS module (www.sparkfun.com/ products/11058). Like most GPS modules, it has TTL Serial output and will send out a burst of messages once a second at 9600 baud.

The messages conform to a standard called National Marine Electronics Association (NMEA). Each message is a string of text, ending with the newline character. The fields of the message are separated by commas. A typical message is shown here:

```
$GPRMC,081019.548,A,5342.6316,N,00239.8728,W,000.0,079.7,110613,,,A*76
```

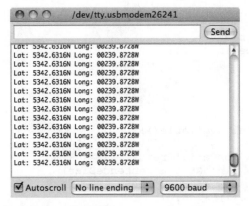

**Figure 10-5**    *GPS readings on an Arduino*

The fields in the example are as follows:

- **$GPRMC**    The sentence type

- **081019.548**    The time (very accurate) and in 24-hour format. 8:10:19.548

- **5342.6316, N**    Latitude × 100, that is, 53.426316 degrees North

- **00239.8728,W**    Longitude × 100, that is, 0.2398728 degrees West

- **000.0**    Speed

- **079.7**    Course 79.7 degrees

- **110613**    Date 11 June 2013

The remaining fields are not relevant to this example.

*NOTE*    *You can find a complete list of the NMEA GPS sentences listed here: http://aprs.gids.nl/nmea/.*

Here is the code for this example:

```
#include <SoftwareSerial.h>

SoftwareSerial gpsSerial(10, 11); // RX, TX (TX not used)
const int sentenceSize = 80;

char sentence[sentenceSize];
```

```
void setup()
{
  Serial.begin(9600);
  gpsSerial.begin(9600);
}

void loop()
{
  static int i = 0;
  if (gpsSerial.available())
  {
    char ch = gpsSerial.read();
    if (ch != '\n' && i < sentenceSize)
    {
      sentence[i] = ch;
      i++;
    }
    else
    {
     sentence[i] = '\0';
     i = 0;
     // Serial.println(sentence);
     displayGPS();
    }
  }
}

void displayGPS()
{
  char field[20];
  getField(field, 0);
  if (strcmp(field, "$GPRMC") == 0)
  {
    Serial.print("Lat: ");
    getField(field, 3);  // number
    Serial.print(field);
    getField(field, 4); // N/S
    Serial.print(field);

    Serial.print(" Long: ");
    getField(field, 5);  // number
    Serial.print(field);
    getField(field, 6);  // E/W
    Serial.println(field);
  }
}
```

```
void getField(char* buffer, int index)
{
  int sentencePos = 0;
  int fieldPos = 0;
  int commaCount = 0;
  while (sentencePos < sentenceSize)
  {
    if (sentence[sentencePos] == ',')
    {
      commaCount ++;
      sentencePos ++;
    }
    if (commaCount == index)
    {
      buffer[fieldPos] = sentence[sentencePos];
      fieldPos ++;
    }
    sentencePos ++;
  }
  buffer[fieldPos] = '\0';
}
```

The sentences coming from the GPS module are of differing lengths, but are all less than 80 characters, so the code uses a buffer variable **sentence** that is filled with the data until an end-of-line marker is read or the buffer is full.

A C null character is placed on the end of the buffer when the whole sentence has been read. This is only so that if you wish, you can "print" the sentence to see the raw data.

The rest of the sketch is concerned with extracting individual fields and formatting the output to be written to the Serial Monitor. The **getField** function helpfully extracts the text from a field at a particular index.

The **displayGPS** function first ignores any sentences that are not of the type **"$GPRMC"** and then extracts the latitude and longitude and hemisphere fields to be displayed.

## Summary

In this chapter, we investigated a few ways to program serial communications between Arduinos, peripherals, and computers.

In the next chapter, we'll turn our attention to an interesting property of the Arduino Leonardo that allows it to emulate USB peripherals such as a keyboard and mouse. We will also look at other aspects of USB Programming.

# 11

# USB Programming

**This chapter** looks at various aspects of using the Arduino with USB. This includes the keyboard and mouse emulation features provided by the Arduino Leonardo and also the reverse process of allowing a USB keyboard or mouse to be connected to a suitably equipped Arduino.

## Keyboard and Mouse Emulation

Three Arduino boards—the Due, the Leonardo, and the Micro, which is based on the Leonardo—can use their USB port to emulate a keyboard or mouse. There are also Arduino-compatible boards like the LeoStick from Freetronics (Figure 11-1) that can perform this trick.

**Figure 11-1** *The LeoStick*

This feature is practically used largely for things like music controllers, giving the Arduino a way to interface with music synthesis and control programs like Ableton Live. You could, for example, build novel musical instruments with Arduino that use accelerometer readings, interrupted beams of light, or pedal boards to control music software.

Some of the sillier applications of these features include pranks where the computer mouse appears to take on a life of its own or the keyboard itself types random letters.

The Arduino Due has two USB ports. Keyboard and mouse emulation takes place on the *native USB port*, and you normally program the Arduino Due using the *programming USB port* (Figure 1-2).

## Keyboard Emulation

The keyboard functions are quite easy to use. They are part of the core language, so there is no library to include. To begin keyboard emulation, simply put the following command in your **startup** function:

```
Keyboard.begin();
```

To have the Arduino "type" something, you can use **print** and **println** commands and the text will appear wherever the cursor is positioned:

Native
USB port

Programming
USB port

**Figure 11-2**   *The Arduino Due's two USB ports*

```
Keyboard.println("It was the best of times.");
```

If you need to use modifier keys, such as typing CTRL-C, then you can use the **press** command:

```
Keyboard.press(KEY_LEFT_CTRL);
Keyboard.press('x');
delay(100);
Keyboard.releaseAll();
```

The **press** command takes a single char as its parameter, and in addition to all the normal characters, a number of constants such as **KEY_LEFT_CTRL** are defined for you to use. Once you issue the **press** command, it is as if the key is held down until the **releaseAll** command is given. You can find a full list of the special keys here: http://arduino.cc/en/Reference/KeyboardModifiers.

*NOTE    When using the keyboard and mouse emulation features, you may encounter difficulty programming the board as it might be trying to type text while you are trying to program it. The trick is to keep the Reset button depressed and only release it when the "uploading" message appears in the status line of the Arduino IDE.*

# Keyboard Emulation Example

The following example automatically types text of your choice (for instance, a password) every time the Arduino is reset:

```
// sketch_11_01_keyboard

char phrase[] = "secretpassword";

void setup()
{
  Keyboard.begin();
  delay(5000);
  Keyboard.println(phrase);
}

void loop()
{
}
```

This example would be better if an external button triggered the typing; if you are using a Mac, the operating system thinks a new keyboard has been attached when you reset the device, which opens a system dialog that you must dismiss before the text is typed.

## Mouse Emulation

Emulating a mouse follows much the same pattern as emulating a keyboard. Indeed there is no reason why you cannot use both in the same sketch.

The first step is to begin emulation:

```
Mouse.begin();
```

You can then move the mouse using **Mouse.move**. The three parameters are the amount to move the x, y, and scroll button in pixels. These numbers can be positive (right or down) or negative (left and up). They are relative to the current mouse position, and as there is no way to get the absolute position of the cursor, this emulation just emulates the mouse that moves the cursor, not the cursor itself.

You can also click the mouse using the **click** command. With no parameters, this command is a simple left button click. You can also optionally supply an argument of **MOUSE_RIGHT** or **MOUSE_MIDDLE**.

If you want to control the duration of a mouse click, then you can use the **Mouse.press** and **Mouse.release** commands. **Mouse.press** takes the same optional arguments as **Mouse.click**. This can be useful if you are, say, making your own mouse from an Arduino and want the button click to be controlled by a switch connected to a digital input on the Arduino. Doing this would allow you to double- or triple-click.

## Mouse Emulation Example

The following example moves the mouse randomly around your screen. To stop the program so you can regain control of your computer, either press and hold down the Reset button or just unplug the board.

```
// sketch_11_02_mouse

void setup()
```

```
{
  Mouse.begin();
}

void loop()
{
  int x = random(61) - 30;
  int y = random(61) - 30;
  Mouse.move(x, y);
  delay(50);
}
```

# USB Host Programming

Whereas the Leonardo, Due, and Micro have the ability to act like a keyboard or mouse, only the Due and the lesser known Arduino Mega ADK have a feature that allows you to connect a USB keyboard or mouse to it so you can use it as an input device. This feature is called *USB Host*, and although only the Due supports it directly, there are third-party USB host shields that you can plug into an Arduino Uno or Leonardo that give you USB Host.

What is more, if you have a wireless keyboard and mouse (not the Bluetooth variety), it should also work if you plug the USB adaptor into the host shield's USB socket. This way you can add wireless remote control to an Arduino.

The USB Host facility is not only restricted to a keyboard and mouse; you can use it for many other USB peripherals, such as video game controllers, cameras, and Bluetooth, and to interface with your Android phone.

## USB Host Shield and Library

The USB Host shield and accompanying libraries have been around for a few years and now support a good range of peripherals. The original shield was developed by Circuits@home (www.circuitsathome.com/). Other compatible USB Host shields are now available from Sparkfun, SainSmart, and probably others. Figure 11-3 shows a Sparkfun USB Host shield attached to an Arduino Uno. Note that at the time of writing, these

**Figure 11-3** *Sparkfun USB Host shield*

boards are not compatible with the Arduino Leonardo, or, in fact, any-thing much more exotic than a Uno. So check before buying.

This particular board has a handy prototyping area to which you can solder your own extra components. An alternative to a shield is to use a board such as the Freetronics USBDroid (Figure 11-4). This board has both a micro USB port for programming the board and a full-size USB socket that you can plug a keyboard or similar into.

If you are using the USBDroid or an unofficial USB Host shield, then you need to use the original USB_Host_Shield library from Circuits@ Home. If you are using the official board, then the USB_Host_Shield_2 library is available with support for more types of devices.

USB programming using the Host Shield Libraries is not easy. The library provides a fairly low-level interface to the USB bus. The sample sketch **sketch_11_03_host_keyboard**, available from the author's web

**Figure 11-4**   *The Freetronics USBDroid*

site (www.simonmonk.org), is an example of connecting a keyboard using a USB Host connection.

This sketch is adapted from one of the example sketches in the examples folder of the USB_Host_Shield library. The code is changed to output keypresses to the Serial Monitor rather than to an LCD screen as in the original example, however.

The sketch (and the original on which it is based) make a useful template for your own code, as they both handle all the keys on the keyboard properly. If you were only interested in the digit or cursor keys, then you could greatly simplify the sketch.

The sketch is too long to list here in full, so instead I will just highlight key areas. You might find it useful to have the sketch loaded while reading the descriptions of the code.

There are three libraries to import:

```
#include <Spi.h>
#include <Max3421e.h>
#include <Usb.h>
```

The **Spi.h** library is required because that is the interface used by the USB Host controller chip. The chip itself is a **Max3421e**, hence, that library, and finally there is another library (**Usb.h**) layered on top of that, which hides some of the complexity of using the chip directly.

After importing the libraries, you'll see a series of constant definitions like this:

```
#define BANG        (0x1E)
```

These are just a different way to define constants in C. It could have also been written like this:

```
const int BANG = 0x1E;
```

MAX3421E and USB objects are then created, and in **setup**, the **powerOn** function of **Max** is invoked.

```
MAX3421E Max;
USB Usb;
```

In the **loop** function, both Max and Usb have their respective **Task** functions called. This triggers the interface to check for USB activity.

```
void loop() {
    Max.Task();
    Usb.Task();
    if( Usb.getUsbTaskState() == USB_STATE_CONFIGURING ) {
//wait for addressing state
        kbd_init();
        Usb.setUsbTaskState( USB_STATE_RUNNING );
    }
    if( Usb.getUsbTaskState() == USB_STATE_RUNNING ) {
//poll the keyboard
        kbd_poll();
    }
}
```

The USB interface will on first startup be in the **USB_STATE_CONFIG-URING** state until the keyboard connection is established when **kbd_init** is called. This function uses an endpoint record structure (**ep_record**) into which the parts of the message are placed to establish a message to begin the link with the keyboard:

```
ep_record[ 0 ] = *( Usb.getDevTableEntry( 0,0 ));
ep_record[ 1 ].MaxPktSize = EP_MAXPKTSIZE;
ep_record[ 1 ].Interval  = EP_POLL;
ep_record[ 1 ].sndToggle = bmSNDTOG0;
ep_record[ 1 ].rcvToggle = bmRCVTOG0;
Usb.setDevTableEntry( 1, ep_record );
/* Configure device */
rcode = Usb.setConf( KBD_ADDR, 0, 1 );
```

After initialization is complete, the most likely state encountered in the main loop is that the keyboard is up and running (**USB_STATE_ RUNNING**), in which case **kbd_poll** is called to check for a keypress on the keyboard.

The key line in **kbd_poll** is

```
rcode = Usb.inTransfer( KBD_ADDR, KBD_EP, 8, buf );
```

This line reads a USB key scancode to see if a key has been pressed. This code is not the same as an ASCII value. The mapping to ASCII takes place in the **HIDtoA** function. This function is the most complex in the sketch, but one that you can easily reuse in your own sketches. You can find a list of scancodes and how they map to ASCII here: www.win.tue.nl/~aeb/ linux/kbd/scancodes-1.html.

One interesting feature of the USB Human Interface Device (HID) protocol used with keyboards is that the SCROLL LOCK and NUM LOCK indicator LEDs can be controlled. This takes place in the **kbd_poll** function in response to the SCROLL LOCK, CAPS LOCK, and NUM LOCK keys being pressed, however, you could write a little sketch like the one in **sketch_11_04_host_scroll_lock** that simply flashes the LEDs.

The key function in this sketch is

```
void toggleLEDs( void )
{
  if (leds == 0) {
    leds = 0b00000111;
  }
```

```
else {
  leds = 0;
}
Usb.setReport( KBD_ADDR, 0, 1, KBD_IF, 0x02, 0, &leds );
}
```

The three least significant bits of the character are flags for the three "lock" LEDs on the keyboard.

## USB Host on the Arduino Due

The Arduino Due hast the ability to act as a built-in USB Host. This feature is, at the time of writing, considered to be "experimental" by the Arduino team. Check the official Arduino documentation (http://arduino.cc/en/Reference/USBHost) for changes to the status of this work or any changes to the way it is used.

The Due does not have a full-size USB socket into which you can directly plug a USB keyboard or mouse. To use such devices, you must get a Micro USB OTG Host Cable like the one attached to the Due pictured in Figure 11-5. In the figure, the USB adapter for a wireless keyboard is attached to the Arduino Due, but a regular USB keyboard would work just fine.

The USB libraries on the Arduino Due are actually a great deal easier to use than the USB Host library and will return the ASCII value of a key that is pressed and not just the USB key scancode. The following example

**Figure 11-5**   *Arduino Due with a Micro USB OTG Host Cable and keyboard*

illustrates interfacing to a keyboard. It simply echoes each keypress in the
Serial Monitor.

```
// sketch_11_05_keyboard_due

#include <KeyboardController.h>

USBHost usb;
KeyboardController keyboard(usb);

void setup()
{
  Serial.begin(9600);
  Serial.println("Program started");
  delay(200);
}

void loop()
{
  usb.Task();
}

// This function intercepts key press
void keyPressed()
{
  char key = keyboard.getKey();
  Serial.write(key);
}
```

The **KeyboardController** library invokes the **keyPressed** function in
the sketch every time a key is pressed. You can also intercept key release
using the **keyReleased** function. To find out which key was pressed, you
must call one of the following functions on the **keyboard** object:

- **getModifiers**  Returns a bit mask for any modifier key that is
  depressed (SHIFT, CTRL, and so on). See http://arduino.cc/en/
  Reference/GetModifiers for the codes.

- **getKey**  Gets the current key as an ASCII value.

- **getOemKey**  Returns the key scancode.

Using a mouse is equally easy and follows a similar pattern to the keyboard controller. The following example writes a letter—*L, R, U,* or *D*—depending on whether the mouse is moved left, right, up, or down:

```
// sketch_11_06_mouse_due

#include <MouseController.h>

USBHost usb;
MouseController mouse(usb);

void setup()
{
  Serial.begin(9600);
  Serial.println("Program started");
  delay(200);
}

void loop()
{
  usb.Task();
}

// This function intercepts mouse movements
void mouseMoved()
{
  int x = mouse.getXChange();
  int y = mouse.getYChange();
  if (x > 50) Serial.print("R");
  if (x < -50) Serial.print("L");
  if (y > 50) Serial.print("D");
  if (y < -50) Serial.print("U");
}
```

As well as the **mouseMoved** function, you can also add the following functions to intercept other mouse events:

- **mouseDragged**   This event is triggered when moving the mouse while holding down the left button.
- **mousePressed**   This event is triggered when a mouse button is pressed and should be followed by a call to **mouse.getButton**, which takes a button name of LEFT_BUTTON, RIGHT_BUTTON,

or MIDDLE_BUTTON as an argument and returns true if it has been pressed.

- **mouseReleased**   This function is the counterpart to **mousePressed** and is used to detect when the mouse has been released.

## Summary

In this chapter, you looked at a few ways to use an Arduino with USB devices.

In the next chapter, we will look at using wired and wireless network connections with an Arduino and learn how to do some network programming as well as make use of the Ethernet and WiFi Arduino shields.

# 12
# Network Programming

**The Internet,** in what has been called the *Internet of Things,* is starting to go beyond browsers and web servers to include Internet-enabled hardware. Printers, home automation devices, and even refrigerators are not only becoming smart, but also being connected to the Internet. And Arduino is at the forefront of DIY Internet devices using either a wired connection to an Ethernet Shield or a WiFi connection. In this chapter, we look at how to program the Arduino to make use of a network connection.

## Networking Hardware

You have a number of choices for connecting your Arduino to the network. You can use an Ethernet Shield with an Arduino Uno or an Arduino with built-in Ethernet hardware, or go for the more expensive, but wireless, WiFi Shield.

## Ethernet Shield

As well as providing an Ethernet connection, the Ethernet Shield (Figure 12-1) also provides a microSD card slot, which you can use to store data (see "Using SD Card Storage" in Chapter 6).

The W5100 chip is used in the official boards; you can also find much lower-cost Ethernet Shields that use the ENC28J60 chipset. These less expensive boards are not compatible with the Ethernet library, however, and are frankly best avoided unless you have more time than budget.

193

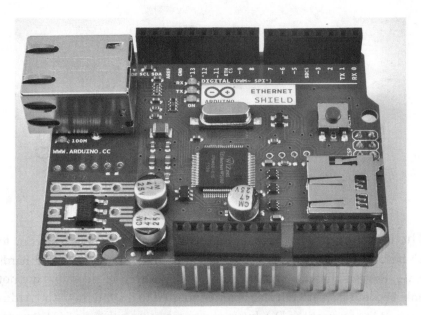

**Figure 12-1** *Ethernet Shield*

## Arduino Ethernet/EtherTen

An alternative to using a separate shield is to buy an Arduino with built-in Ethernet capability. The official version is the Arduino Ethernet, but a worthy and Uno-compatible third-party board is the EtherTen produced by Freetronics (www.freetronics.com). This board is shown in Figure 12-2.

Combining everything onto one board makes a lot of sense when building a networked Arduino project. The Arduino Ethernet can also be fitted with a Power over Ethernet (PoE) adapter that, with a separate PoE injector, allows the board to be powered from an Ethernet lead, reducing the wires needed for the Arduino to be just a single Ethernet lead. The EtherTen board comes already configured to use PoE. For more information on using PoE with an EtherTen, see www.doctormonk .com/2012/01/power-over-ethernet-poe.html.

## Arduino and WiFi

The problem with an Ethernet Internet connection is, of course, that it requires a wire. If you want your Arduino to connect to the Internet or to

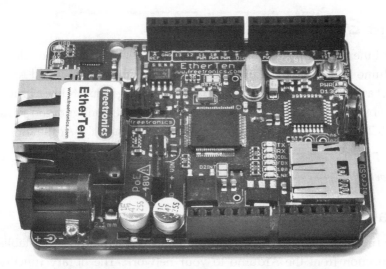

**Figure 12-2**   *An EtherTen board*

a network and operate wirelessly, then you need a WiFi Shield (Figure 12-3). These are somewhat expensive, but third-party alternatives are available such as the Sparkfun WiFly shield (https://www.sparkfun.com/products/9954).

**Figure 12-3**   *An Arduino WiFi Shield*

# The Ethernet Library

The Ethernet library has undergone a major revision since the release of Arduino 1.0 in 2011. In addition to allowing an Ethernet-equipped Arduino to act as either a web server or a web client (sending requests like a browser), the library also handles things like Dynamic Host Configuration Protocol (DHCP), which automatically assigns an IP address to the Arduino.

*NOTE*   *The official Arduino documentation on the Ethernet library is actually very good: http://arduino.cc/en/reference/ethernet.*

## Making a Connection

The first step, before any communication can take place, is to establish a connection from the Arduino to your network. The library function is called **Ethernet.begin()**. You can manually specify the connection settings for the board using the following syntax:

```
Ethernet.begin(mac, ip, dns, gateway, subnet)
```

Let's look at each of these parameters in turn:

- **mac**   The mac address of the network card (I'll explain this in a moment.)
- **ip**   The IP address of the board (You have to select one acceptable to your network.)
- **dns**   The IP address for a Domain Name Server (DNS)
- **gateway**   The IP address for the Internet gateway (your home hub)
- **subnet**   The subnet mask

This syntax looks a little daunting unless you are used to manual network administration. Fortunately, all the parameters except **mac** are optional, and 90 percent of the time, you will either specify **mac** and **ip** or, most likely, just the **mac** on its own. All the other settings are taken care of automatically.

The MAC, or Media Access Control, address is a unique identifier for the network interface; in other words, it's the address for the Ethernet Shield or for whatever is providing the network interface for the Arduino.

This strange-looking code only has to be unique for your network. You'll usually find this number printed on the back of your Arduino Ethernet Shield or WiFi Shield (Figure 12-4) or on the box packaging. If you are using an older board that does not have a MAC address, then you can simply make one up. However, do not use the same made-up number more than once on your network.

You can also create a network connection using DHCP so the IP address is allocated dynamically; use this code:

```
#include <SPI.h>
#include <Ethernet.h>

byte mac[] = { 0xDE, 0xAD, 0xBE, 0xEF, 0xFE, 0xED };
void setup()
{
  Ethernet.begin(mac);
}
```

**Figure 12-4**  *Mac address sticker on a WiFi Shield*

If you want to fix the IP address, which would be desirable if you wanted to run a web server on the Arduino, then you would use code like this:

```
#include <SPI.h>
#include <Ethernet.h>
byte mac[] = { 0xDE, 0xAD, 0xBE, 0xEF, 0xFE, 0xED };
byte ip[] = { 10, 0, 1, 200 };

void setup()
{
  Ethernet.begin(mac, ip);
}
```

You need to ensure that the IP address you use is acceptable to your network. If you do not specify an IP address and use DHCP, then **Ethernet.begin** will return 1 if a connection is made and an IP address allocated; otherwise, it returns a 0. You can incorporate a test in which you make the connection and use the **localIP** function to retrieve the IP address allocated to the Arduino. The following example performs this test and reports the status to the Serial Monitor. This is a full sketch that you can try for yourself. But before you do, remember to change the MAC address in the code to match that of your interface board.

```
// sketch_12_01_dhcp

#include <SPI.h>
#include <Ethernet.h>


byte mac[] = { 0x00, 0xAA, 0xBB, 0xCC, 0xDE, 0x02 };

void setup()
{
  Serial.begin(9600);
  while (!Serial){}; // for Leonardo compatibility

  if (Ethernet.begin(mac))
  {
    Serial.println(Ethernet.localIP());
  }
```

```
  else
  {
    Serial.println("Could not connect to network");
  }
}

void loop()
{
}
```

# Setting Up a Web Server

The project "Physical Web Server," described later in this chapter, illustrates the code structure of a web server sketch. In this section, we'll look at the available web server functions.

The **EthernetServer** class contains most of the functions that you need for web serving. Having established a connection to the network, starting a web server requires two further steps. First, you need to create a new server object, specifying the port that the server should be listening on. This declaration appears in the sketch before the **setup** function.

```
EthernetServer server = EthernetServer(80);
```

Web pages are usually served on port 80. So if you start the web server on port 80, you will not need to add a port number to any URL that connects to the server.

Second, to actually start the server, you use the following command in your **setup** function:

```
server.begin();
```

This function starts the server, and it will now be waiting for someone with a browser to load the web page that it is serving. This action is detected in the **loop** function of your sketch using the **available** function. This function returns either null (if there are no requests to service) or an **EthernetClient** object. This object is rather confusingly also used when making outgoing requests from the Arduino to web servers. In either case, **EthernetClient** represents the connection between a web server and a browser.

Having retrieved this object, you can then read the incoming request using **read** and you can write HTML to it using the **write**, **print**, and **println** functions. Once you've finished writing the HTML to the client, you need to call stop on the client object to end the session. I explain how to do this in "Physical Web Server" later in this chapter.

# Making Requests

In addition to having the Arduino act as a web server, you can also have it act like a web browser, issuing HTTP requests to a remote web server, which may be on your own network or on the Internet.

When making web requests from the Arduino, you first establish a network connection in just the same way that you did in the previous section for the web server, but instead of creating an **EthernetServer** object, you create an **EthernetClient** object:

```
EthernetClient client;
```

You do not need to do anything more with the client object until you want to send a web request. Then you write something like this:

```
if (client.connect("api.openweathermap.org", 80))
  {
    client.println("GET /data/2.5/weather?q=Manchester,uk HTTP/1.0");
    client.println();
    while (client.connected())
    {
      while (client.available())
      {
        Serial.write(client.read());
      }
    }
    client.stop();
  }
```

The **connect** function returns true if the connection is successful. The two **client.println** commands are responsible for requesting the desired page from the web server. The two nested **while** loops then read data as long as the client is connected and data is available.

It may look tempting to combine the two **while** loops, with a condition of **client.available() && client.connected()**, but combining them is not quite the same as treating them separately, as the data may not be available continuously from the web server because of connection speed and so on. The outer **while** loop keeps the request alive, fetching the data.

This approach is "blocking" (the Arduino will not do anything else until the request is complete). If this is not acceptable for your project, you can include code to check for other conditions inside the inner **while** loop.

# Ethernet Examples

The following two examples serve to illustrate the use of the Ethernet library in practical settings. Between the two of them, the examples cover most things that you are likely to want to do with a networked Arduino.

## Physical Web Server

This first example illustrates perhaps the most likely web-related use of an Arduino. In it, the Arduino acts as a web server. Browsers connecting to the Arduino web server not only see readings from the analog inputs, but visitors can also press buttons on the web page to change the digital outputs (Figure 12-5).

This example is actually a great way to interface an Arduino with a smartphone or tablet computer, as a device only has to have the most basic of browsers on it to be able to send requests to the Arduino. The sketch for this example (**sketch_12_02_server**) is some 172 lines long, so rather than list it in full here, I encourage you to load it in the Arduino IDE for reference as I walk you through it.

The first part of the sketch is pretty standard for a network sketch. The libraries are imported, and both **EthernetServer** and **EthernetClient** objects are defined.

The variables in the next section perform various roles:

```
const int numPins = 5;
int pins[] = {3, 4, 5, 6, 7};
int pinState[] = {0, 0, 0, 0, 0};
char line1[100];
char buffer[100];
```

The constant **numPins** defines the size of the arrays **pins** and **pinState**. The **pinState** array is used to remember whether the particular output pin

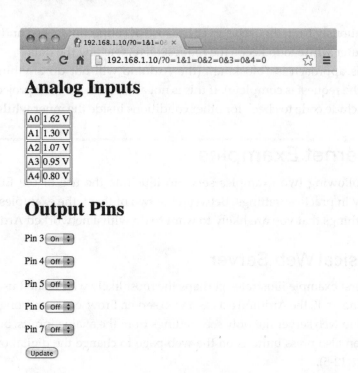

**Figure 12-5** *Physical web server interface*

is HIGH or LOW. The **setup** function declares all the pins in the **pins** array
to be outputs. It also establishes the network connection in the same way
as in the earlier examples. Finally, the **line1** and **buffer** character arrays
hold the first line of the HTTP request and subsequent lines, respectively.

Here is the **loop** function:

```
void loop()
{
  client = server.available();
  if (client)
  {
    if (client.connected())
    {
      readHeader();
```

```
    if (! pageNameIs("/"))
    {
      client.stop();
      return;
    }
    client.println(F("HTTP/1.1 200 OK"));
    client.println(F("Content-Type: text/html"));
    client.println();

    sendBody();
    client.stop();
  }
 }
}
```

The **loop** function checks to see if there are any requests from browsers waiting to be processed. If there is a request and there is a connection, then the **readHeader** function is called. You'll find the code for this toward the end of the sketch. The **readHeader** function reads the contents of the request header into a buffer (line1) and then skips over the remaining lines of the header. This is required so you have access to the page name (requested by the browser) as well as any request parameters.

Note that because the sketch has a fair amount of text to send to the Serial Monitor and network, I use the **F** function to store the character arrays in flash memory (see Chapter 6).

Having read the header, the **pageNameIs** function (again near the end of the file) is called to ensure the page being requested is the root page (/). If it is not the root page, then the request is ignored. This is important because many browsers automatically send a request to a server to find an icon for the website. You don't want this request being confused with other requests to the server.

Now you need to generate a response in the form of a header and some HTML to be returned to the browser for display. The **sendHeader** function generates an "OK" response to indicate to the browser that the request is valid. The **sendBody** function, shown here, is a lot more complicated:

```
void sendBody()
{
```

```
    client.println(F("<html><body>"));
    sendAnalogReadings();
    client.println(F("<h1>Output Pins</h1>"));
    client.println(F("<form method='GET'>"));
    setValuesFromParams();
    setPinStates();
    sendHTMLforPins();
    client.println(F("<input type='submit' value='Update'/>"));
    client.println(F("</form>"));
    client.println(F("</body></html>"));
}
```

This function prints the basic template of the HTML page, relying on a number of helper functions to break the code down into more manageable chunks. The first of these is **sendAnalogReadings**:

```
void sendAnalogReadings()
{
    client.println(F("<h1>Analog Inputs</h1>"));
    client.println(F("<table border='1'>"));
    for (int i = 0; i < 5; i++)
    {
        int reading = analogRead(i);
        client.print(F("<tr><td>A")); client.print(i);
        client.print(F("</td><td>"));
        client.print((float) reading / 205.0);
        client.println(F(" V</td></tr>"));
    }
    client.println("</table>");
}
```

This loops over each of the analog inputs, reading the value and writing out an HTML table containing all the readings as voltages.

You may have noticed that the **sendBody** function also calls **setValues-FromParams** and **setPinStates**. The first of these uses the function **value-OfParam** to set the **pinStates** variable containing the states of the output pins HIGH or LOW depending on the value of the request parameters that were contained in the request header:

```
int valueOfParam(char param)
{
    for (int i = 0; i < strlen(line1); i++)
    {
        if (line1[i] == param && line1[i+1] == '=')
```

```
    {
      return (line1[i+2] - '0');
    }
  }
  return 0;
}
```

The **valueOfParam** function expects the request parameter to be a single digit. You can see what these parameters look like if you run the example and browse to the page and press **Update**. The URL string will then change to include the parameters and look something like this:

```
192.168.1.10/?0=1&1=0&2=0&3=0&4=0
```

The parameters start after the **?** and take the form *X=Y*, separated by **&**. The part before the = is the parameter name (in this case, a digit from 0 to 4) and the part after the = is its value, which is 1 for on and 0 for off. To make life easy for yourself, these request parameters must be only a single character or, in this case, a single digit. The **setPinStates** function then transfers the state of the output pins held in the **pinStates** array to the actual output pins themselves.

Let's return to the **sendBody** function for a moment. The next thing that you need to send is the HTML for the collection of drop-down lists for each output. You need the values of **True** or **False** in the list to be set to agree with the current state of the output. You accomplish this by adding the text "selected" to the value that agrees with the value for that pin in the **pinStates** array.

All the HTML generated for the output pins is contained within a form, so when a visitor presses the **Update** button, a new request to this page with the appropriate request parameters to set the outputs is generated. At this point, let's look at the HTML code that is generated for the page:

```
<html><body>
<h1>Analog Inputs</h1>
<table border='1'>
<tr><td>A0</td><td>0.58 V</td></tr>
<tr><td>A1</td><td>0.63 V</td></tr>
<tr><td>A2</td><td>0.60 V</td></tr>
<tr><td>A3</td><td>0.65 V</td></tr>
<tr><td>A4</td><td>0.60 V</td></tr>
```

```
</table>
<h1>Output Pins</h1>

<form method='GET'>
<p>Pin 3<select name='0'>
<option value='0'>Off</option>
<option value='1' selected>On</option>
</select></p>
<p>Pin 4<select name='1'>
<option value='0' selected>Off</option>
<option value='1'>On</option>
</select></p>
<p>Pin 5<select name='2'>
<option value='0' selected>Off</option>
<option value='1'>On</option>
</select></p>
<p>Pin 6<select name='3'>
<option value='0' selected>Off</option>
<option value='1'>On</option>
</select></p>
<p>Pin 7<select name='4'>
<option value='0' selected>Off</option>
<option value='1'>On</option>
</select></p>
<input type='submit' value='Update'/>
</form>
</body></html>
```

You can see this using your browser's View Source feature.

# Using a JSON Web Service

To illustrate sending a web request from an Arduino to a website, I'll use a web service that returns data about the weather in a particular location. It reports a short description of the weather to the Serial Monitor (Figure 12-6). The sketch sends the request once during startup, but the example could easily be changed to check every hour and display the result on a 16×2 LCD display.

The sketch for this example is quite short, just 45 lines of code (**sketch_12_03_web_request**). Most of the interesting code is in the function **hitWebPage**:

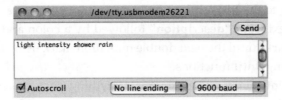

**Figure 12-6** *Retrieving weather information from a web service*

```
void hitWebPage()
{
  if (client.connect("api.openweathermap.org", 80))
  {
    client.println("GET /data/2.5/weather?q=Manchester,uk HTTP/1.0");
    client.println();
    while (client.connected())
    {
      if (client.available())
      {
        client.findUntil("description\":\"", "\0");
        String description = client.readStringUntil('\"');
        Serial.println(description);
      }
    }
    client.stop();
  }
}
```

The first step is to get the client to connect to the server on port 80. If this is successful, then the page request header is written to the server:

```
client.println("GET /data/2.5/weather?q=Manchester,uk HTTP/1.0");
```

The extra **println** is needed to mark the end of the request header and trigger a response from the server.

To wait for the connection, the **if** statement inside the **while** loop detects when data is available to be read. Reading the data stream directly avoids the need to capture all of the data into memory. The data is in JSON format:

```
{"coord":{"lon":-2.23743,"lat":53.480949},
"sys":{"country":"GB","sunrise":1371094771,
"sunset":1371155927},"weather":[{"id":520,"main":"Rain",
"description":"light intensity shower rain","icon":"09d"}],
"base":"global stations","main":{"temp":284.87,"pressure":1009,
"humidity":87,"temp_min":283.15,"temp_max":285.93},
"wind":{"speed":5.1,"deg":270},"rain":{"1h":0.83},
"clouds":{"all":40},"dt":1371135000,"id":2643123,
"name":"Manchester","cod":200}
```

Returning to the **hitWebPage** function, we are going to extract the section of the text from **"description"** followed by a colon and then double quotation marks until the next double quotation mark using the **findUntil** and **readStringUntil** functions.

The **findUntil** function just ignores everything from the server until the matching string is found. The **readStringUntil** function then reads all the subsequent text until the double quote character.

# The WiFi Library

As you might expect, the WiFi library is quite similar to the Ethernet library. If you substitute **WiFi** for **Ethernet**, **WiFiServer** for **Ethernet-Server**, and **WiFiClient** for **EthernetClient**, then everything else in your sketch can pretty much stay the same.

## Making a Connection

The main differences between the WiFi and Ethernet libraries are in how a connection is established.

First, you need to import the WiFi library:

```
#include <SPI.h>
#include <WiFi.h>
```

To establish a connection, use the **WiFi.begin** command, supplying it with the name of your wireless network and your password.

```
WiFi.begin("MY-NETWORK-NAME", "mypassword");
```

The WiFi example that follows in "WiFi Example" illustrates the other differences that you need to be aware of.

## WiFi Specific Functions

The WiFi library has some extra WiFi-specific functions that you can use. These functions are summarized in Table 12-1.

You can find full documentation for the WiFi library here: http://arduino.cc/en/Reference/WiFi.

| Function | Description |
|---|---|
| WiFi.config | Allows you to set static IP addresses and DNS and gateway addresses for the WiFi adapter |
| WiFi.SSID | Returns a string containing the SSID (that is, the wireless network name) |
| WiFi.BSSID | Returns a byte array with the MAC address of the router that the WiFi Shield is connected to |
| WiFi.RSSI | Returns a **long** containing the signal strength |
| WiFi.encryptionType | Returns a number code for the encryption type |
| WiFi.scanNetworks | Returns the number of networks found, but no other information about them |
| WiFi.macAddress | Places the MAC address of the WiFi adapter into a 6-byte array passed as its parameter |

**Table 12-1**  *WiFi Specific Features*

# WiFi Example

For the example, I modified **sketch_12_02_server** to work with a WiFi Shield. You can find the code in **sketch_12_04_server_wifi**. Rather than repeat the whole example, I will just highlight the changes from the original version.

First, to make the connection to a wireless access point, you need to specify the name of the wireless network and its password:

```
char ssid[] = "My network name";  // your network SSID (name)
char pass[] = "mypassword";       // your network password
```

You also need to change the names of the classes for the server and client from **EthernetServer** and **EthernetClient** to **WiFiServer** and **WiFiClient**:

```
WiFiServer server(80);
WiFiClient client;
```

You still need to specify port 80 when defining the server.

The next difference between the two shields is at the point where the connection starts. In this case, you must use

```
WiFi.begin(ssid, pass);
```

The remainder of the code is almost exactly the same as the Ethernet code. You will find a **delay(1)** command in **loop** before the client is stopped, which gives the client time to finish reading before the communication is closed. You don't need this in the Ethernet version. You'll also notice that I combined some of the **client.print** calls into fewer calls of bigger strings. This speeds up the communication as the WiFi Shield deals with sending small strings quite inefficiently. However, be aware that the strings in an individual **client.print** or **client.println** cannot be longer than 90 bytes or they will not be sent.

The WiFi version of this program is considerably slower than the Ethernet version, taking up to 45 seconds to load. The firmware on the WiFi Shield can be updated, and if in the future the Arduino team improves the efficiency of the WiFi Shield, then it may be worth updating the firmware. Look for instructions for this on the WiFi Shield web page: http://arduino.cc/en/Main/ArduinoWiFiShield.

# Summary

In this chapter, you looked at a variety of ways to connect your Arduino to a network and then make it do something, using both Ethernet and Wi-Fi Shields. You have also learned how to use an Arduino as both a web server and a web client.

In the next chapter, you'll learn about Digital Signal Processing (DSP) with the Arduino.

# 13

# Digital Signal Processing

**The Arduino** is capable of fairly rudimentary signal processing. This chapter discusses a variety of techniques, from conditioning a signal from an analog input using software rather than external electronics to calculating the relative magnitude of various frequencies in a signal using a Fourier Transform.

## Introducing Digital Signal Processing

When you take readings from a sensor, you are measuring a signal. It is common to visualize that signal as a line (usually wavy) moving from the left of the page to the right over time. This is how electrical signals are viewed on an oscilloscope. The *y*-axis is the *amplitude* of the signal (its strength) and the *x*-axis is time. Figure 13-1 shows a signal in the form of music, captured over a period of just 1/4 of a second using an oscilloscope.

You can see some repeating patterns in the signal. The frequency at which these patterns recur is called the *frequency*. This is measured in Hertz (abbreviated to Hz). A signal of 1 Hz repeats itself once every second. A signal of 10 Hz, 10 times per second. Looking at the left of Figure 13-1, you see a signal that repeats itself roughly every 0.6 of a square. As each square represents 25 milliseconds, with the settings used on the oscilloscope, the frequency of that part of the signal is $1/(0.6{\times}0.025) = 67$ Hz. If you were to zoom in using a shorter time span, you would see that many other sound component frequencies mixed in there as well. Unless a signal is pure sine

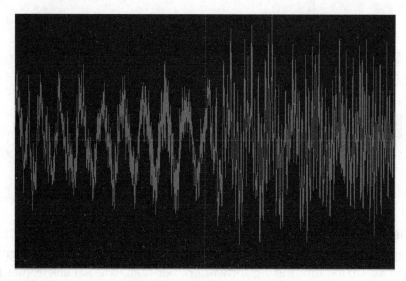

**Figure 13-1**   *A signal from a musical source*

wave (like the one shown later in Figure 13-5), then it will always comprise a whole load of frequencies.

You could try to capture the signal shown in Figure 13-1 using one of the Arduino's analog inputs. This is called *digitization* because you are making the analog signal digital. To do this, you have to be able to take samples fast enough to get a good reproduction of the original signal.

The essence of Digital Signal Processing (DSP) is to digitize a signal using an analog-to-digital converter (ADC), manipulate it in some way, and then generate an analog output signal using a digital-to-analog converter (DAC). Most modern audio equipment, MP3 players, and cell phones use DSP, which provides equalization settings that allow you to control the relative power of the high or low frequencies in a piece of music. Sometimes, however, you don't need the output to be a version of the input; you simply need to use DSP techniques to remove unwanted noise from a signal to get more accurate readings from a sensor.

In general, Arduinos are not the ideal devices for DSP. They cannot capture analog signals particularly fast, and their digital output is limited to PWM. The exception to this is the Arduino Due, which, as well as having lots of ADCs also has a fast processor and two true DACs. Therefore, the Due's hardware is sufficiently good enough to stand a fighting chance of digitizing a stereo audio signal and doing something with it.

# Averaging Readings

When reading from sensors, you often find that you can get better results by taking a number of readings and then averaging them. One way to do this is to use a circular buffer (Figure 13-2).

Using a circular buffer arrangement, as each new reading is taken, it is added to the buffer at the current index position. When the last index position is filled, the index position is set back to zero and the old readings start being overwritten. In this way, you always keep the last $N$ readings, where $N$ is the size of the buffer.

The following example code implements a circular buffer:

```
// sketch_13_01_averaging

const int samplePin = A1;

const int bufferSize = 10;
int buffer[bufferSize];
int index;

void setup()
{
  Serial.begin(9600);
}

void loop()
{
  int reading = analogRead(samplePin);
  addReading(reading);
  Serial.println(average());
  delay(1000);
}

void addReading(int reading)
{
  buffer[index] = reading;
```

```
    index++;
    if (index >= bufferSize) index = 0;
}

int average()
{
    long sum = 0;
    for (int i = 0; i < bufferSize; i++)
    {
        sum += buffer[i];
    }
    return (int)(sum / bufferSize);
}
```

This approach produces invalid averages until the buffer has been filled. In practice, this need not be a problem as you can just ensure that you take a buffer full of readings before you start requesting the average.

Notice that the **average** function uses a **long** to contain the sum of the readings. Using a **long** is essential if the buffer is long enough to exceed the maximum **int** value of about 32,000. Note that the return value can still be an **int** as the average will be within the range of the individual readings.

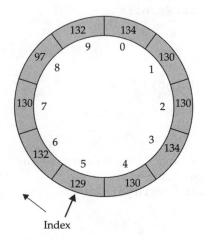

**Figure 13-2**  *A circular buffer*

# An Introduction to Filtering

As I discussed in the section "Introducing Digital Signal Processing," any signal is usually comprised of a wide range of different component frequencies. At times, you may want to ignore some of these frequencies, in which case you need to use filtering.

The most common type of filtering with an Arduino is probably *low-pass filtering*. Say you have a light sensor and you are trying to detect the overall light level and how it changes from minute to minute, for instance, to detect when it is dark enough to turn on a light. But you want to eliminate higher frequency events such as a hand momentarily passing near the sensor or the sensor being illuminated by artificial light that actually flickers considerably at the line frequency (60 Hz if you live in the United States). If you are only interested in the very slow-moving part of the signal, then you need a low-pass filter.

For the opposite effect, if you want to respond to fast-moving events but ignore the longer trend, you need a *high-pass filter*.

Returning to the line frequency interference problem, if, for example, you are interested in frequencies above and below the 60 Hz noise, then simply cutting off the low frequencies may not be an option. For that, you may want to use a *band stop filter* that just removes the component of the signal at 60 Hz or, more likely, all frequencies from 59 to 61 Hz.

# Creating a Simple Low-Pass Filter

Maintaining a buffer of readings is often unnecessary if all you really want to do is smooth out the signal. Such filtering can be thought of as low-pass filtering because you are rejecting high-frequency rapid signal changing and are interested in the overall trend. You use filters like this in sensors such as accelerometers that are sensitive to high-frequency changes in the signal that you may not be interested in if you simply want to know the angle something is tilted to.

A simple-to-code and useful technique for accomplishing this relies on retaining a kind of running average between readings. This running average comprises a proportion of the current running average and a proportion of the new reading:

$$smoothedValue_n = (alpha \times smoothedValue_{n-1}) + ((1 - alpha) \times reading_n)$$

*Alpha* is a constant between 0 and 1. The higher the value of *alpha*, the greater the smoothing effect.

This makes it sound more complicated than it is, however. The following code shows how easy it is to implement:

```
// sketch_13_02_simple_smoothing
const int samplePin = A1;
const float alpha = 0.9;

void setup()
{
  Serial.begin(9600);
}

void loop()
{
  static float smoothedValue = 0.0;
  float newReading = (float)analogRead(samplePin);
  smoothedValue = (alpha * smoothedValue) +
    ((1 - alpha) * newReading);
  Serial.print(newReading); Serial.print(",");
  Serial.println(smoothedValue);
  delay(1000);
}
```

By copying and pasting the output of the Serial Monitor into a spreadsheet and then charting the result, you can see how well the smoothing is performing. Figure 13-3 shows the result of the previous code, with a short wire stuck into the top of A1 to pick up some electrical interference.

You can see how it takes a while for the smoothed value to catch up. If you were to increase *alpha* to, say, 0.95, then the smoothing would be even more pronounced. Plotting the data written to the Serial Monitor is a great way to make sure the smoothing that you are applying to your signal is what you need.

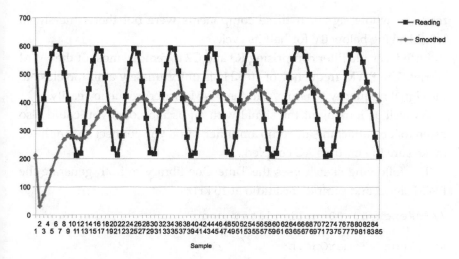

**Figure 13-3** *Plotting smoothed values*

# Arduino Uno DSP

Figure 13-4 shows how you can wire up an Arduino so an audio signal is fed into A0 and a PWM (10 kHz) output signal is generated. I used a smartphone app as the signal generator, and I connected the headphone output of the phone to the Arduino, as shown in Figure 13-4.

*CAUTION*  *Be warned: connecting your phone in this way probably voids its warranty and could destroy your phone.*

The input from the signal generator is biased using C1, R1, and R2; therefore, the oscillation is about the midpoint of 2.5V, so the ADC can

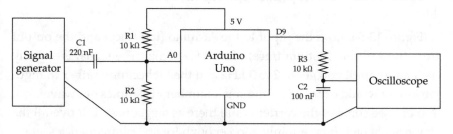

**Figure 13-4** *Using an Arduino Uno for DSP*

read the whole signal. If these components were not there, the signal would swing below 0V for half its cycle.

I used a crude filter comprising R3 and C2 to remove most of the PWM carrier. The PWM frequency of 10 KHz is unfortunately a bit too close to the signal frequency to remove all the PWM carrier frequency easily.

As well as looking at the signal with an oscilloscope, you could also listen to it by attaching an audio amplifier, but if you connect an amplifier, make sure the input is AC coupled.

The following sketch uses the TimerOne library to both generate the PWM signal and sample the audio at 10 kHz:

```
// sketch_13_03_null_filter_uno

#include <TimerOne.h>

const int analogInPin = A0;
const int analogOutPin = 9;

void setup()
{
  Timer1.attachInterrupt(sample);
  Timer1.pwm(analogOutPin, 0, 100);
}

void loop()
{
}

void sample()
{
  int raw = analogRead(analogInPin);
  Timer1.setPwmDuty(analogOutPin, raw);
}
```

Figure 13-5 shows the input to the Arduino (top trace) and the output from the Arduino (bottom trace) of a 1 kHz signal. The signal is actually not bad up until you get to 2 to 3 kHz and then it becomes rather triangular, as you would expect with the small number of samples per waveform. You can see some of the carrier is still there as jaggedness, but overall the shape is not bad. It is certainly good enough for speech frequencies.

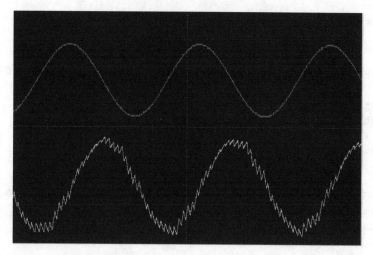

**Figure 13-5**  *Arduino Uno signal reproduction with a signal of 1 kHz*

# Arduino Due DSP

Now we can carry out the same experiment using an Arduino Due at a much higher sample rate. The code used for the Uno in the previous section is of no use with the Due, which cannot use the TimerOne library and has a different architecture.

The Due analog inputs operate at 3.3V, so be sure to connect the top of R1 to 3.3V and *not* 5V. Because the Due has an analog output, you can dispense with the low-pass R3 and C2 filter and connect the oscilloscope directly to the DAC0 pin. Figure 13-6 shows the connections for the Due.

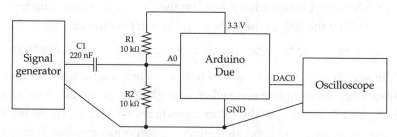

**Figure 13-6**  *Using an Arduino Due for DSP*

The following sketch uses a sample rate of 100 kHz!

```
// sketch_13_04_null_filter_due

const long samplePeriod = 10L; // micro seconds

const int analogInPin = A0;
const int analogOutPin = DAC0;

void setup()
{
  // http://www.djerickson.com/arduino/
  REG_ADC_MR = (REG_ADC_MR & 0xFFF0FFFF) | 0x00020000;
  analogWriteResolution(8);
  analogReadResolution(8);
}

void loop()
{
  static long lastSampleTime = 0;
  long timeNow = micros();
  if (timeNow > lastSampleTime + samplePeriod)
  {
    int raw = analogRead(analogInPin);
    analogWrite(analogOutPin, raw);
    lastSampleTime = timeNow;
  }
}
```

Unlike the other Arduino boards, the Arduino Due allows the resolution of both the ADC and DAC to be set. To keep things simple and fast, these are both set to 8 bits.

The following line speeds up ADC on the Due by manipulating register values. Follow the link in the code for more information on this trick.

```
REG_ADC_MR = (REG_ADC_MR & 0xFFF0FFFF) | 0x00020000;
```

The sketch uses the **micros** function to control the sample frequency, only running the sampling code when enough microseconds have elapsed.

Figure 13-7 shows how the setup reproduces a 5 kHz input signal. You can see the steps in the generated signal corresponding to the 20 samples per waveform you would expect from a 100 kHz sample rate.

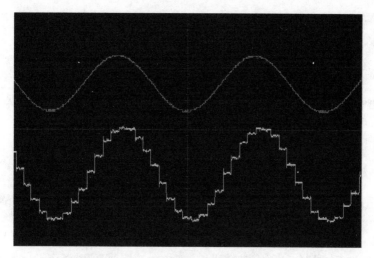

**Figure 13-7**  *Arduino Due signal reproduction with a signal of 5 kHz*

# Filter Code Generation

If you are looking at more advanced filtering, there is a useful online code generator that allows you to design a filter and then cut and paste lines of code that it generates into your Arduino sketch. You will find this code here: http://www.schwietering.com/jayduino/filtuino/.

Frankly, the alternative is whole lot of painful math!

Figure 13-8 shows the interface to the filter generator. The bottom half of the screen shows the generated code, and shortly I will show you how you can incorporate this into an Arduino sketch.

You have a bewildering array of options for the type of filter to be generated. The example shown in Figure 13-4 is a band stop filter designed to reduce the amplitude of the signal at frequencies between 1 kHz and 1.5 kHz. Starting at the top row, the settings for this are "Butterworth," "band stop," and "1st order." *Butterworth* refers to the filter design, from its original analog electronics design (http://en.wikipedia.org/wiki/Butterworth_filter). The Butterworth is a good all round design and a good default.

I also selected the option "1st order." Changing this to a higher number will increase both the number of previous samples that need to be stored

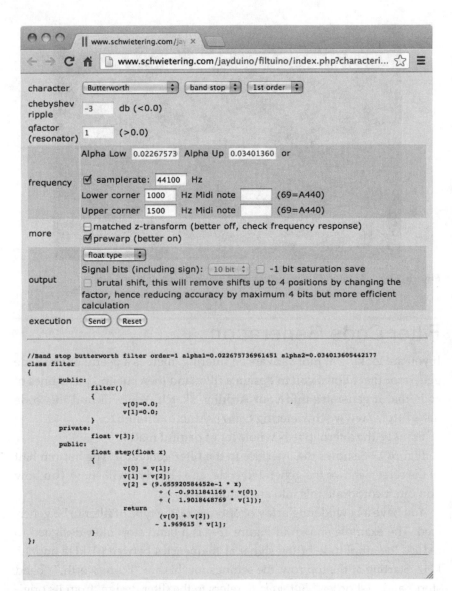

**Figure 13-8**   *Filter code generator for Arduino*

and also the steepness of the cutoff of the unwanted frequencies. For this example, "1st order" is fine. The higher the order, the more calculations to perform and you may find that you need to reduce the sample rate for the Arduino to keep up.

Then you see some disabled fields that relate to other types of filter, until you come to "samplerate." Samplerate is the frequency at which the data will be sampled and also the frequency at which the generated code will be called to apply the filtering to the signal.

Next, I specified the upper and lower frequencies. You can enter these as either a frequency in Hz or a MIDI note.

The "more" section provides a couple more options and even tells you how to set them for best results. The "output" section allows you to specify the type to use for the array of values that are used to do the filtering. I set this to "float." Finally, I clicked Send to generate the code.

To test this, you can modify the "null filter" example that you ran on the Due. The full sketch can be found in **sketch_13_05_band_stop_due**.

The first step is to copy and paste the generated code into the basic "null filter" example just after the constant definitions. It is also a good idea to paste the URL from the generator as a comment line, so if you want to go back and modify the filter code, you'll have the parameters you used last time preset in the user interface. The generated code encapsulates all the filter code into a class. You'll meet classes again in Chapter 15. But, for now, you can treat it as a black box that will do filtering.

After the pasted code, you need to add the following line:

```
filter f;
```

Now you need to modify the **loop** function, so that instead of simply outputting the input, the Arduino outputs the filtered value:

```
void loop()
{
   static long lastSampleTime = 0;
   long timeNow = micros();
   if (timeNow > lastSampleTime + samplePeriod)
   {
     int raw = analogRead(analogInPin);

     float filtered = f.step(raw);

     analogWrite(analogOutPin, (int)filtered);
     lastSampleTime = timeNow;
   }
}
```

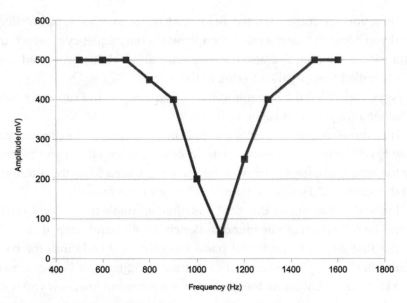

**Figure 13-9** *Frequency response of an Arduino band stop filter*

Making the filtered signal is as easy as supplying the raw reading from the analog input as argument to the function **f.step**. The value returned is the filtered value, which can be converted to an **int** before being written by the DAC.

Looking at the **step** function, you can see that the filter code keeps a history of three previous values along with the new value. There is some shuffling up of values and then scaling of values by factors to produce a return value. Isn't math wonderful?

Figure 13-9 shows the result of this filtering. A signal generator was used to inject different frequency signals and the output amplitude (measured using the oscilloscope) recorded in a spreadsheet and then plotted in a chart.

# The Fourier Transform

The Fourier Transform is a useful tool for analyzing the frequencies in a signal. As you recall from the introduction to this chapter, signals can be thought of as being made up of varying amounts of sine waves at different frequencies. You have probably seen frequency analyzer displays on musical equipment or on the visualization utilities in your favorite MP3 playing software.

These analyzers display as bar charts. The vertical bars represent the relative strengths of different bands of frequencies, with the low-frequency bass notes over on the left and the high-frequency bands on the right.

Figure 13-10 shows how the same signal can be viewed both as a single wavy line (called the *time domain*) and a set of strengths of the signal in a set of frequency bands (called the *frequency domain*).

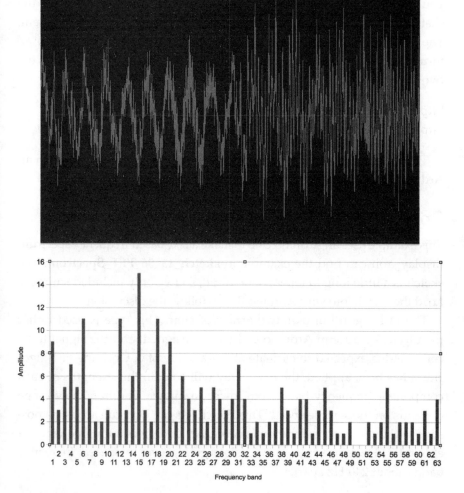

**Figure 13-10**   *A signal in time and frequency domains*

The algorithm for calculating the frequency domain from time domain signal data is called the *Fast Fourier Transform* or *FFT*. The calculations involved in FFTs use complex numbers and, unless you are really into math, are not for the faint of heart.

Fortunately for us, clever people are often happy to share their code. You can download a function that will perform the FFT algorithm for you. The sample code I used is not organized into a library; it is supplied as a C header and C++ implementation file (.h and .cpp, respectively). To use it, you can just place the two files into your sketch folder. These are in the sketches that accompany the book, so you do not need to download them. The code originally appeared in a post on the Arduino Forum (http://forum.arduino.cc/index.php/topic,38153.0.html). You can also find the two files, with other examples of the algorithm, at these websites:

https://code.google.com/p/arduino-integer-fft/
https://github.com/slytown/arduino-spectrum-analyzer/

The following two examples illustrate how to use code running on an Arduino Uno to sample an audio signal.

## Spectrum Analyzer Example

This example uses an Arduino Uno to make a text-based frequency spectrum display. You can find the example in **sketch_13_06_FFT_Spectrum**. The sketch is a little long to repeat here in full, so I've only included snippets. Load the sketch into your Arduino IDE to follow the discussion.

The FFT algorithm uses two arrays of **char**. This type is used rather than **byte**, because in Arduino C, **byte** is unsigned, and the signal to be converted is expected to oscillate about a value of 0. Once the FFT algorithm has been applied, the **data** array will contain the strengths of each component frequency band from lower to higher. The frequency range depends on the sample speed. This sketch lets the Uno run as fast as possible without any consideration for accuracy and gives a top frequency of about 15 kHz, since there are 63 slots giving evenly spaced frequency bands about 240 Hz apart.

To make the analog conversion as fast as possible and get a decent sample rate, use the trick discussed in Chapter 4 to increase the speed of the ADC. This accounts for these two lines in **setup**.

```
ADCSRA &= ~PS_128;   // remove prescale of 128
ADCSRA |= PS_16;     // add prescale of 16 (1MHz)
```

The main **loop** doesn't contain much code:

```
void loop()
{
  sampleWindowFull();
  fix_fft(data, im, 7, 0);
  updateData();

  showSpectrum();
}
```

The function **sampleWindowFull** samples a time window of 128 samples worth of data. I'll discuss this in a moment. The FFT algorithm is then applied. The parameter of 7 is the base 2 logarithm of the number of samples. This value will always be 7. The parameter of 0 is an inverse flag, which will also always be set to 0 for false. After the FFT algorithm has been applied, there is a further step to update the values in the arrays. Finally **showSpectrum** is called to display the frequency data.

The function **sampleWindowFull** reads 128 analog values and assumes that the signal is biased to 2.5V, so that by subtracting 512 from the reading, the signal will swing both positive and negative. This is then scaled by the constant **GAIN** to add a little amplification for weak signals. The 10-bit reading is then converted into an 8-bit value to fit into the **char** array by dividing it by 4. The **im** array containing the imaginary component of the signal is set to 0. This is part of the inner workings of the algorithm; if you want to find out more about this, see http://en.wikipedia.org/wiki/Fast_Fourier_transform.

```
void sampleWindowFull()
{
  for (int i = 0; i < 128; i++)
  {
    int val = (analogRead(analogPin) - 512) * GAIN;
```

```
        data[i] = val / 4;
        im[i] = 0;
    }
}
```

The **updateData** function calculates the amplitude of each frequency slot. The strength of the signal is the hypotenuse of the right-angle triangle whose other sides are the real and imaginary parts of the signal (Pythagoras's Theorem in action!).

```
void updateData()
{
    for (int i = 0; i < 64; i++)
    {
        data[i] = sqrt(data[i] * data[i] + im[i] * im[i]);
    }
}
```

To display the data, it is written to the Serial Monitor, which places the whole data set on one line, with commas between the values. The first value is ignored, as this contains the DC component of the signal and is not usually of interest.

You could, for example, use the **data** array to control the height of graphical bars on an LCD display. To connect a signal (say, the audio output of an MP3 player), you would need the same type of arrangement as shown previously in Figure 13-4 so the signal is biased around 2.5V.

## Frequency Measurement Example

This second example uses an Arduino Uno to display the approximate frequency of a signal in the Serial Monitor (**sketch_13_07_FFT_Freq**). Most of the code is the same as for the previous example. The main difference is that once the **data** array has been calculated, the index position of its highest value is used to calculate an estimate of the frequency. The **loop** then displays this value in the Serial Monitor.

```
int findF()
{
    int maxValue = 0;
    int maxIndex = 0;
    for (int i = 1; i < 64; i++)
```

```
{
    int p = data[i];
    if (p > maxValue)
    {
      maxValue = p;
      maxIndex = i;
    }
  }
  int f = maxIndex * 240;
  return f;
}
```

## Summary

DSP is a complex subject, and there are many books devoted just to this topic alone. This complexity means that, by necessity, I've only touched on what are hopefully the more useful and possible techniques that you might try with an Arduino.

In the next chapter, we turn our attention to dealing with the problem of wanting to do more than one thing at a time with the Arduino. This is a particular problem for those of us used to programming larger machines, where multiple simultaneous threads of execution are the norm.

# 14

# Managing with One Process

**Programmers coming** to Arduino from a background in programming large systems often cite the lack of multithreading and concurrency in Arduino as a deficiency. In this chapter, I'll try to set the record straight and show you how to embrace the single-thread model of embedded systems.

## Making the Transition from Big Programming

Arduino has attracted many enthusiasts, including me, who have spent years in the software industry and are used to teams of dozens of people contributing to a huge software effort, with all the related problems of managing the ensuing complexity. For us, the ability to write a few lines of code and have something interesting and physical happen almost immediately, without large amounts of engineering, is the perfect antidote to big software.

It does, however, mean that we often look for things in Arduino that we are used to seeing in our day jobs. When moving from the big development world to the miniature world of Arduino, one of the first adjustments we need to make is to the very simplicity of writing for Arduino. To develop a large system without the benefit of Test Driven Development, version control, and some kind of Agile process to follow is reckless. On the other hand, a large Arduino project may be only 200 lines of code written by one person. If that person is an experienced software developer, he or she can simply keep the details in mind without needing any of the usual accoutrements of development.

So stop fretting about version control, design patterns, writing unit tests, and having a refactoring IDE and just embrace the joyous simplicity of Arduino.

## Why You Don't Need Threads

If you are old enough to have programmed home computers in BASIC, then you remember that "doing one thing at a time" is simply how computers operate. In BASIC, if a game required a number of sprites to be moved apparently simultaneously, then you had to be smart and include a main loop that moved each sprite a little bit.

This mindset is a good one to have for Arduino programming. Rather than multiple threads each being responsible for one of the sprites, a single execution thread does a little bit of everything in turn, without "blocking" on any one thing.

Aside from multicore computers, essentially a computer only genuinely does one thing at once. The rest of the time, the operating system switches the processor's attention among the numerous processes running on the computer. On the Arduino, with a limited need to do more than one thing at a time, you can code it yourself, as there is no operating system.

## Setup and Loop

It is no accident that the two functions you must write for any sketch are **setup** and **loop**. The fact that **loop** repeats over and over again, indicates why you should not really allow loop to block. Your code should wiz through loop and around again before you know it.

## Sense Then Act

Most Arduino projects contain an element of needing to control something. Therefore, the contents of a loop often:

- Check if buttons are pressed or a sensor threshold has been exceeded.
- Perform a relevant action.

A simple example of this would be a push switch that, when pressed, toggles LED flashing on and off.

The following example illustrates this. As you shall see later, however, the limitations imposed by having to wait while the LED flashes are sometimes not acceptable.

```
// sketch_14_01_flashing_1

const int ledPin = 13;
const int switchPin = 5;
const int period = 1000;

boolean flashing = false;

void setup()
{
  pinMode(ledPin, OUTPUT);
  pinMode(switchPin, INPUT_PULLUP);
}

void loop()
{
  if (digitalRead(switchPin) == LOW)
  {
    flashing = ! flashing;
  }
  if (flashing)
  {
    digitalWrite(ledPin, HIGH);
    delay(period);
    digitalWrite(ledPin, LOW);
    delay(period);
  }
}
```

The problem with this code is that you can only check that the button has been pressed once the blinking has finished. If a button is pressed while the blinking is in progress, it won't register. This may not be important to the operation of the sketch, but if it is important to register every button press, then you need to make sure the loop does not have any delays in it. In fact, once the flashing is triggered, the Arduino spends most of its time blinking and there is only a tiny window in which the button press can be registered.

The example in the next section solves this problem.

## Pause Without Blocking

You can rewrite the previous sketch to avoid using **delay**:

```
// sketch_14_02_flashing_2

const int ledPin = 13;
const int switchPin = 5;
const int period = 1000;

boolean flashing = false;
long lastChangeTime = 0;
int ledState = LOW;

void setup()
{
  pinMode(ledPin, OUTPUT);
  pinMode(switchPin, INPUT_PULLUP);
}

void loop()
{
  if (digitalRead(switchPin) == LOW)
  {
    flashing = ! flashing;
    // and turn the LED off
    if (! flashing)
    {
      digitalWrite(ledPin, LOW);
    }
  }
  long now = millis();
```

```
if (flashing && now > lastChangeTime + period)
{
  ledState = ! ledState;
  digitalWrite(ledPin, ledState);
  lastChangeTime = now;
}
}
```

In this sketch, I have added two new variables: **lastChangeTime** and **ledState**. The **lastChangeTime** variable records the last time the LED was toggled between on and off, and the **ledState** variable contains that on/off state, so when it needs to be toggled, you know what the LED's current state is.

The loop now contains no delays. The first part of the loop checks for a button press, and if a button is pressed, it toggles the flashing state. The extra **if** statement, shown next, is simply a nice refinement that turns the LED off if the button press has caused flashing to be turned off. Otherwise, the LED might be left on, even though flashing has been canceled.

```
if (! flashing)
{
  digitalWrite(ledPin, LOW);
}
```

The second part of the loop finds the current **millis()** count and then compares this with the value in **lastChangeTime** with **period** added to it. This means that the code inside the **if** will only be run if more than **period** milliseconds has elapsed.

The **ledState** variable is then toggled and the digital output set accordingly. The value in **now** is then copied to **lastChangeTime** so the code can wait for the next **period** to elapse before being activated again.

# The Timer Library

The "Pause Without Blocking" approach of the previous section has been generalized into a library that allows you to schedule repeating events using **millis**. Despite its name, the library has nothing to do with the hardware timers on the device and will, therefore, work just fine on most Arduino boards.

You can download the library from http://playground.arduino.cc//Code/Timer.

Using this library simplifies the code, as you can see here:

```
// sketch_14_03_flashing_3
#include <Timer.h>

const int ledPin = 13;
const int switchPin = 5;
const int period = 1000;

boolean flashing = false;
int ledState = LOW;
Timer t;

void setup()
{
  pinMode(ledPin, OUTPUT);
  pinMode(switchPin, INPUT_PULLUP);
  t.every(period, flashIfRequired);
}

void loop()
{
  if (digitalRead(switchPin) == LOW)
  {
    flashing = ! flashing;
    if (! flashing)
    {
      digitalWrite(ledPin, LOW);
    }
  }
  t.update();
}

void flashIfRequired()
{
  if (flashing)
  {
    ledState = ! ledState;
    digitalWrite(ledPin, ledState);
  }
}
```

To use this library, you define a timer, in this case called **t**, and then within your **setup** function you specify a function that calls periodically using:

```
t.every(period, flashIfRequired);
```

You then place the following line in your **loop** function:

```
t.update();
```

Every time the **update** function is called, **millis** checks when any of the timed events need to be actioned, and if they do, it calls the linked function (in this case **flashIfRequired**).

The Timer library also has a number of other utility functions; for more information on the library, see the link at the beginning of this section.

# Summary

In this chapter, you learned how to allow multiple things to appear to happen at the same time on an Arduino, without using multiple threads. This is simply a matter of adjusting your mindset to the constraints imposed by your favorite little microcontroller board.

In the final chapter of this book, you will learn how to share your code creations with the Arduino community by creating and publishing Arduino libraries.

# 15

# Writing Libraries

Sooner or later you will create something really good that you think other people could make use of. This is the time to wrap up the code in a library and release it to the world. This chapter shows you how.

## When to Make a Library

Creating an Arduino library is not an activity restricted to Arduino developers; any Arduino user can create a library. If it's useful, then much praise will flow in the developer's direction. No one sells libraries—that would be counter to the values of the Arduino community. Libraries should be released as open source as a way to help your fellow Arduino enthusiasts.

Perhaps the most useful Arduino libraries are those that are developed to provide an interface to a specific piece of hardware. They often greatly simplify the process of using the hardware and, in particular, unraveling some complex protocol. There is no reason why more than one person should have to go through the pain of working out how some obscure bit of hardware works, and thanks to the Internet, if you publish a helpful library, people will generally find it.

*TIP* *The application programmer interface (API) is the set of functions that the library user will include in his or her sketch. When designing the API, always ask yourself this question: "What does the user actually care about?" The low-level implementation details should be hidden as much as possible. In the example developed in "Library Example (TEA5767 Radio)," I'll discuss this issue further.*

# Using Classes and Methods

Although the sketch writer generally has the impression that he or she is writing in C and using a fairly conservative set of C features, in actual fact, the sketch writer is using C++. Arduino sketches are based on C++, the object-oriented extension to the C language. This language uses the concepts of *classes* of objects that group together information about the object (its data) and also functions that apply to the data. These functions look like regular functions but when associated with a particular class are referred to as *methods*. What is more, methods can be declared to be public, in which case anyone can use them, or private, in which case they are only accessible to other methods inside the class.

The reason I am telling you all this is that extension writing is one of the few Arduino activities in which using classes is the norm. The class is a great way to wrap up everything into a kind of module. The "private"/"public" distinction is also a good way to ensure that when you are designing the API, you are always thinking of how the sketch writer will want to interact with the library (the public) rather than how it works (the private).

As you work through the example that follows, you'll see how to use a class.

# Library Example (TEA5767 Radio)

To illustrate how to write an Arduino Library, I'll wrap up some code that you first met back in Chapter 7 for the TEA5767 FM radio receiver. The sketch is simple and barely justifies a library, but nonetheless, it serves as a good example.

The following are the stages in the process:

1. Define the API.
2. Write the header file.
3. Write the implementation file.
4. Write the keywords file.
5. Make some examples.

In terms of files and folders, a library comprises a folder, whose name should match the name of the library class. In this case, I'll call the library and class **TEA5767Radio**. Within that folder, there should be two files: **TEA5767Radio.h** and **TEA5767Radio.cpp**.

Optionally, you may also have a file named **keywords.txt** and a folder called examples, containing example sketches that use the library. The folder structure for this example library is shown in Figure 15-1.

Probably the easiest way to work on the library is directly in your Arduino libraries folder, where you have been installing other third-party libraries. You can edit the files directly in this folder. The Arduino IDE will only register that the library exists once you restart it, but after that any changes to the contents of the files will be picked up automatically when you compile the project.

You can see the original sketch on which this library is based in **sketch_07_01_I2C_TEA5767**, and you can download the finished library from http://playground.arduino.cc//Main/TEA5767Radio.

# Define the API

The first step is to define the interface that people will use.

If you have used a few libraries, you've probably noticed that they generally follow one of two patterns. The simplest is exemplified by the Narcoleptic library. To use this library, you simply include the library and then access its methods by prefixing the method name with Narcoleptic, as shown here:

```
#include <Narcoleptic.h>
// then somewhere in your code
Narcoleptic.delay(500);
```

**Figure 15-1**  *Folder structure of the example project*

This pattern is also used in the Serial library. If there will only ever be one of the things that the library represents, then this pattern is the right one to use. However, if it is possible that there will be more than one, then you want to use a different approach. Because you might want to have more than one radio receiver attached to an Arduino at a time, this particular example falls into this second category.

For these situations, the pattern is similar to that used in the Software-Serial library. Because you might have lots of soft-serial ports at the same time, you create named instances of the SoftwareSerial library using a syntax like this:

```
#include <SoftwareSerial.h>
SoftwareSerial mySerial(10, 11); // RX, TX
```

When you want to use that particular serial port (the one using pins 10 and 11), you create a name for it—in this instance, "mySerial"—and then you can then write things like the following:

```
mySerial.begin(9600);
mySerial.println("Hello World");
```

Without worrying about how you will write the code, let's define how you would like to be able to use the code in a sketch.

After importing the library, you want to be able to create a new "radio," name it, and specify which I2C address it runs on. To make life really easy, you have two options: one where it defaults to the normal port of 0x60 and a second where you specify the port:

```
#include <TEA5767Radio>
TEA5767Radio radio = TEA5767Radio();
// or TEA5767Radio radio = TEA5767Radio(0x60);
```

Because this is an FM radio, what you really need to do is set the frequency, so you need to write something like this in your code.

```
radio.setFrequency(93.0);
```

The number here is the frequency in MHz. It is in the form that the sketch writer would like it in, not in the strange unsigned **int** format that is sent to the module. You want to hide the hard work and wrap it up in the library.

That's all there is to the design in this case. Now we'll write some code.

# Write the Header File

The code for a library is split across more than one file—generally just two files. One file is called the "header" file and has the extension ".h." This file is the one you reference from your sketch using **#include**. The header file does not contain any actual program code; it simply defines the names of the classes and methods in the library. Here is the header file for the example library:

```
#include <Wire.h>

#ifndef TEA5767Radio_h
#define TEA5767Radio_h

class TEA5767Radio
{
private:
  int _address;
public:
  TEA5767Radio();
  TEA5767Radio(int address);
  void setFrequency(float frequency);
};

#endif
```

The **#ifndef** command prevents the library from being imported more than once and is standard practice for header files.

You then include the class definition, which has a private section just containing a variable called **_address**. This variable contains the I2C address for the device.

The public section contains the two functions for creating a radio object— one that allows an address to be specified and one that does not and will, therefore, use the default. The **setFrequncy** function is also listed as public.

# Write the Implementation File

The code that actually implements the functions defined in the header file is contained in the file **TEA5767Radio.cpp**:

```
#include <Arduino.h>
#include <TEA5767Radio.h>
```

```
TEA5767Radio::TEA5767Radio(int address)
{
  _address = address;
}

TEA5767Radio::TEA5767Radio()
{
  _address = 0x60;
}

void TEA5767Radio::setFrequency(float frequency)
{
        unsigned int frequencyB = 4 * (frequency *
          1000000 + 225000) / 32768;
        byte frequencyH = frequencyB >> 8;
        byte frequencyL = frequencyB & 0XFF;
        Wire.beginTransmission(_address);
        Wire.write(frequencyH);
        Wire.write(frequencyL);
        Wire.write(0xB0);
        Wire.write(0x10);
        Wire.write(0x00);
        Wire.endTransmission();
        delay(100);
}
```

The two methods responsible for creating a new radio both simply set the value of **_address** to either the default I2C address of 0x60 or the "address" parameter supplied. The **setFrequency** method is almost identical to the original sketch (**sketch_07_01_I2C_TEA5767**), except that the following line uses the value of the **_address** variable to make the I2C connection:

```
Wire.beginTransmission(_address);
```

## Write the Keywords File

The folder containing the library should also contain a file called **keywords.txt**. This file is not essential; the library will still work if you do not create this file. The file allows the Arduino IDE to color-code any keywords for the library. Our example library only has two keywords: the name of the library itself (**TEA5767Radio**) and **setFrequency**. The keyword file for the library can contain comments on lines that start with a **#**. The keyword file for this library is shown here:

```
######################################
# Syntax Coloring Map for TEA5767Radio
######################################
######################################
# Datatypes (KEYWORD1)
######################################
TEA5767Radio    KEYWORD1
######################################
# Methods and Functions (KEYWORD2)
######################################
setFrequency    KEYWORD2
```

The keywords should be specified as KEYWORD1 or KEYWORD2, although version 1.4 of the Arduino IDE colors both levels of keyword as orange.

# Make the Examples Folder

If you create a folder named **examples** within the folder for the library, then any sketches in the folder will automatically be registered by the Arduino IDE when it starts, so you can access them from the **Examples** menu under the name of the library. The examples sketch can just be a regular sketch, but one that is saved in the folder for the library. The example using this library is listed here:

```
#include <Wire.h>
#include <TEA5767Radio.h>

TEA5767Radio radio = TEA5767Radio();

void setup()
{
  Wire.begin();
  radio.setFrequency(93.0); // pick your own frequency
}

void loop()
{}
```

## Testing the Library

To test the library, you can just run the example sketch that uses the library. Unless you are very lucky (or careful), the library will not work the first time you compile it, so read the error messages that appear in the information area at the bottom of the Arduino IDE.

## Releasing the Library

Having created a library, you need to release it to the community. Perhaps the best way to make sure that people find it is to create an entry on the publicly editable wiki at http://playground.arduino.cc//Main/LibraryList. You can also host the zip file here, although some people prefer to host the library on GitHub, Google Code, or some other hosting platform, and then they just place a link to the code on the wiki.

If you want to upload your library to the Arduino Playground, follow these steps:

1. Test the library to make sure it works as expected.

2. Create a zip file of the library folder and give it the same name as the library class with an extension of .zip.

3. Register yourself as a user on www.arduino.cc.

4. Add an entry on the Arduino Playground wiki—http:// playground.arduino.cc//Main/LibraryList—that describes the library and explains how to use it. The easiest way to do this is look at an entry for another library and copy the appropriate bit of wiki markup. Create a link using, for example, [[TEA5767Radio]] to set a placeholder for a new page that will appear on the library list with a "?" next to it. Clicking the link will create the new page and open the wiki editor. Copy and adapt the wiki code from another library (perhaps from "TEA5767Radio").

5. To upload the zip file of the library, you need to include a tab like this in the wiki markup: **Attach:TEA5767Radio.zip**. After the page has been saved, clicking the download link allows you to specify a zip file to upload onto the wiki server.

# Summary

Creating a library can be very rewarding. Before creating one, however, always search in case someone else has already created the library for you.

The nature of a book like this is that, inevitably, it cannot cover everything that the reader wants to know. But I do hope it has helped you with some of the more common advanced Arduino programming topics.

You can follow me on Twitter as @simonmonk2, and you will find more information about this book and my other books on my website at www .simonmonk.org.

# A

# Parts

As this is a book essentially about programming, not many parts are referenced in this book. This appendix lists the parts that were used, however, and some possible suppliers.

## Arduino Boards

Such is the popularity of Arduino that the common boards like the Uno and Leonardo are readily available. For the less common boards, take a look at Adafruit and SparkFun in the United States as well as CPC in the United Kingdom. Their websites are listed in the "Suppliers" section at the end of this appendix.

## Shields

Adafruit and SparkFun both stock a wide range of the official Arduino shields as well as their own takes on some of the shields. You will also find interesting and low-cost shields and Arduino clones at Seeed Studio.

Shields referenced in the book are listed here. Product codes are in parentheses after the supplier names.

| Shield | Chapter | Sources |
|---|---|---|
| USB Host Shield | 11 | SparkFun (DEV-09947) |
| Ethernet Shield | 12 | Most suppliers |
| WiFi Shield | 12 | Most suppliers |

# Components and Modules

Specific components and modules used as examples in the book are listed here. Product codes are in parentheses after the supplier names.

| Module | Chapter | Sources |
|---|---|---|
| TEA5767 FM receiver module | 7 | eBay |
| LED Backpack Matrix display | 7 | Adafruit (902) |
| DS1307 RTC module | 7 | Adafruit (264) |
| DS18B20 temperature sensor | 8 | Adafruit (374), SparkFun (SEN-00245) |
| MCP3008 8-channel ADC | 9 | Adafruit (856) |
| Venus GPS module | 10 | SparkFun (GPS-11058) |

# Suppliers

There are many suppliers of electronics and Arduino-related parts. A few are listed here:

| Supplier | URL | Notes |
|---|---|---|
| Adafruit | www.adafruit.com | Adafruit products also stocked worldwide by local suppliers. |
| SparkFun | www.sparkfun.com | SparkFun products also stocked worldwide by local suppliers |
| Seeed Studio | www.seeedstudio.com | Unusual and low-cost modules and Arduino clones |
| Mouser Electronics | www.mouser.com | Offer a vast range of all types of electronic parts |
| RadioShack | www.radioshack.com | Stock the more common Arduino parts |
| Digi-Key | www.digikey.com | A vast range of all types of electronic parts |
| CPC | http://cpc.farnell.com | UK supplier with a large range of parts |
| Maplins | www.maplin.co.uk | UK supplier stocking the more common Arduino parts, along with local stores |

# Index